Synthesis Lectures on Technology and Health

The series publishes state-of-the-art short books on transformative technologies for health, wellness, and independent living. Our scope of publishing in the expanding health tech field includes:

- Technology in support of active and healthy living and aging
- Digital technologies for health- and social-care improvement
- Diagnostic, screening, and tracking tools
- Assistive and rehabilitative technologies

The series includes a subseries of books published in partnership with Canada's AGE-WELL that specifically addresses their 8 AgeTech Challenge Areas. Each lecture introduces the context in which the technology is used—wellness, health, medicine, special needs, or other contexts. Authors present and explain the technology and review promising applications and opportunities as well as limitations and challenges. They include material on their own work while surveying the broader landscape of related research, development, and impact.

Alex Mihailidis

The Age of AgeTech

Essays to Inspire, Change, and Challenge

Springer

Alex Mihailidis
University of Toronto
Toronto, ON, Canada

AGE-WELL NCE
Toronto, ON, Canada

ISSN 2771-7054 ISSN 2771-7070 (electronic)
Synthesis Lectures on Technology and Health
ISBN 978-3-032-04885-1 ISBN 978-3-032-04886-8 (eBook)
https://doi.org/10.1007/978-3-032-04886-8

This Springer imprint is published by the registered company Springer Nature Switzerland AG
The registered company address is: Gewerbestrasse 11, 6330 Cham, Switzerland

Dedication

This collection of essays on AgeTech is the culmination of countless hours of research, mentorship, collaboration, and moments of inspiration, none of which would have been possible without the unwavering support, encouragement, and wisdom of the extraordinary people in my life. I want to express my deepest gratitude to every individual who has played a part in shaping the trajectory of my work. Without your guidance and belief in me, these pages would still be a distant dream rather than a written testament of the ideas and experiences I've gathered.

First, this collection is dedicated to my colleagues and collaborators, who challenged me and insisted on pushing the envelope just a little further every time we tackled a new project. You were the ones who kept me honest by asking the tough questions when Ipresented half-baked ideas and who celebrated with me when a risk finally paid off. Through countless whiteboard brainstorming sessions, conference calls, and frantic last-minute proposal submissions, we forged a bond that goes far beyond professional courtesy. We became a family, one that laughs together when the data doesn't make sense and one that rallies around the shared mission of improving the lives of older adults. The spirit of AgeTech is alive in each of you, and I hope these essays reflect the collective resilience, innovation, and heart you embody daily.

Yet, none of this would have been possible without the watchful guidance and mentorship of one remarkable individual: Dr. Geoff Fernie. From the earliest days when my ideas were more enthusiasm than clarity, Dr. Fernie saw something in me that I could not yet articulate for myself. He recognized my passion for leveraging technology to support aging populations and taught me how to channel that passion into meaningful research questions and rigorous methodology. He was unflinching in his criticisms but equally generous with his praise. Geoff, you taught me that real innovation stands at the intersection of creativity and evidence-based practice, and you showed me that the best mentors are unafraid to let their protégés stumble because it's in those stumbles that we learn to run. If these essays possess any scientific merit or practical relevance, it is primarily thanks to your lessons about discipline, curiosity, and empathy for the people behind the data. Thank you for guiding me with the perfect blend of firmness and compassion. Be brave!!

An immeasurable debt is owed to my students and trainees, whose influence on me is far more profound than I had expected. You arrived in my lab eager to learn and grow, but I quickly discovered I was learning from you. Your fresh perspectives, unfiltered questions, bold experiments—and yes, even your missteps—kept me curious and reminded me of why I fell in love with this work in the first place. Each new idea you pursued and every research problem you tackled reaffirmed just how vast the field of AgeTech is and how many uncharted territories await those brave enough to explore. I have always believed that we, professors, mentors, researchers, truly know we've succeeded when our students eclipse us, achieving feats we could only imagine in our wildest dreams. I'm proud to say that I've seen that happen with many of you. Your accomplishments are the most profound measure of my success, and your boundless enthusiasm continues to motivate and humble me daily.

I would be remiss if I didn't recognize the transformative impact of the AGE-WELL team, whose support I cannot possibly thank enough. Together, we have advanced the AgeTech field in ways I would never have accomplished alone. You believed in the power of translating research into real-world impact, never losing sight of the older adults and caregivers at the heart of our mission. Special thanks go to Bridgette Murphy, who, from our first meeting in 2014, believed in all of my wild ideas and did everything possible to turn them into reality. She protected me from the day-to-day challenges that threatened to derail our progress, always had my back, and was the quiet force ensuring everything worked. Without Bridgette's devotion, leadership, and unwavering confidence in AgeTech's potential, so much of what we accomplished would not have been possible.

I can't reflect on this journey without celebrating the love, support, and sacrifices made by my wife, Christine. Christine, your devotion and encouragement have been my bedrock through every iteration of these essays, every long day at the office, and every conference trip that took me away from home. We have seen good times, moments of great triumph, and endured stressful, uncertain times when life demanded more from us than we thought we could manage. Through it all, my excitement when a new grant was approved, my frustration when data didn't pan out, my exhaustion when deadlines piled up, you remained steadfast, supportive, and ready with a comforting word and a nudge when needed. You understood that pursuing AgeTech research was more than a job, it was a calling. If these essays stand for anything, I hope they stand for the enduring power of patience and partnership. Thank you for staying by my side through every challenge, listening to every half-formed idea, and reminding me that everything is figureoutable.

I also dedicate these pages to my children, Bennett and Drew, who are a constant source of joy, perspective, and laughter. You remind me every day why the work I do matters. As I strive to create a world that respects and cares for its older adults, I hope you see that empathy, determination, and creativity can meaningfully change people's lives. Your presence is a testament to the necessity of building a future we can all be proud of—one where technology is harnessed not for mere convenience but for genuine human betterment. Bennett and Drew, I hope you are as proud of me as I am of you. Your curiosity and kindness reassure me that the future is in capable, compassionate hands.

Finally, these essays are more than just an academic exercise to anyone reading this collection. They are a roadmap of the relationships, challenges, triumphs, and reflections that have shaped my perspective on AgeTech and its promise to transform how we care for each other. It is a challenge to push boundaries, stand up for those whose voices aren't always heard, and imagine a world where technology amplifies our humanity rather than eclipsing it.

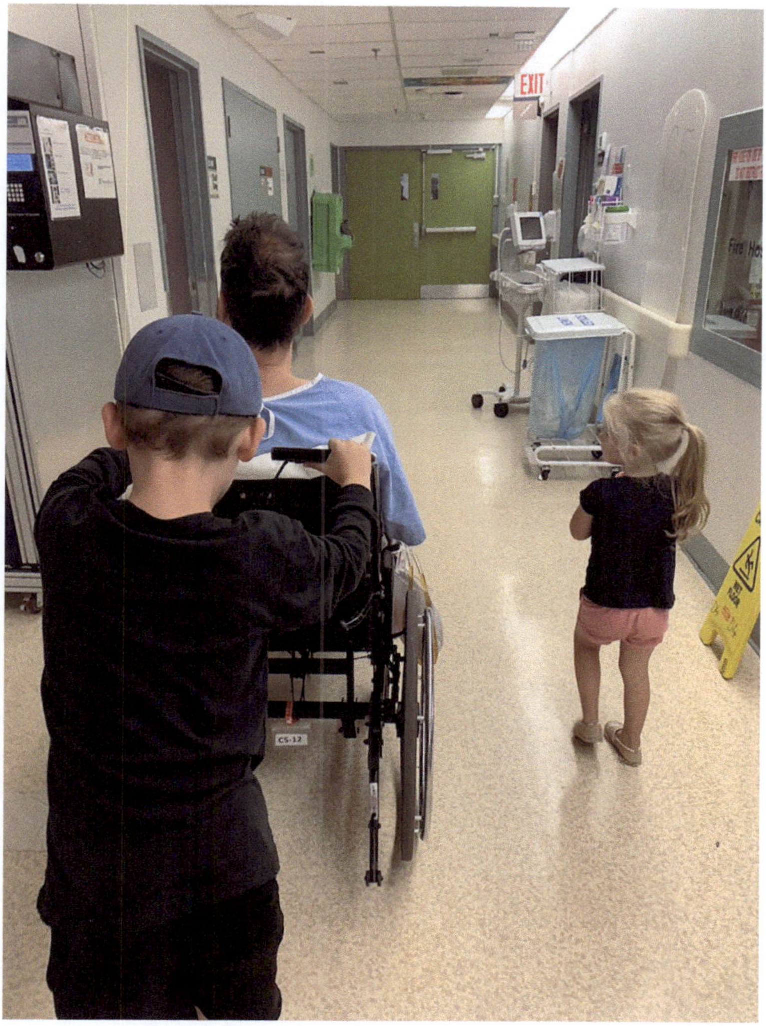

In my darkest hour,
Your little hands guided me into the light

Acknowledgments A very special thank you to my longtime research manager, Bing Ye for keeping my research lab and team running smoothly over all these years and for editing the text of this book.

Competing Interests The author has no competing interests to declare that are relevant to the content of this manuscript.

Contents

1 **Why Did I Write These Essays? More Importantly, Why Should You Read Them?** ... 1
- 1.1 Setting the Scene: A Personal Reflection ... 1
- 1.2 Overview of Essays ... 2
- 1.3 Why Should You Care? ... 4

2 **If You're Not Aging, You're Dead** ... 5
- 2.1 The Science Behind Aging ... 5
- 2.2 How Aging Feels? ... 6
- 2.3 A Global Phenomenon ... 6
- 2.4 Aging in Canada ... 6
- 2.5 Why Aging Matters to Everyone? ... 7
- 2.6 The Need for Innovation and Policy Change ... 7
- 2.7 Opportunities Amid Challenges ... 7
- 2.8 A Shared Responsibility ... 8
- References ... 8

3 **Addressing the Imbalance: The Urgent Need for Increased Funding in Aging Research** ... 9
- 3.1 Introduction ... 9
- 3.2 The Demographic Shift Toward an Aging Population ... 9
- 3.3 The Impact on Healthcare Systems ... 10
- 3.4 Funding Disparities: Aging Research Versus Other Areas ... 10
- 3.5 Healthcare Expenditures: Aging Versus Disease-Specific Costs ... 11
  - 3.5.1 Potential Economic and Health Benefits of Increased Aging Research Funding ... 11
- 3.6 Barriers to Increased Funding for Aging Research ... 11
- 3.7 The Case for Increased Investment in Aging Research ... 12
- 3.8 Aging as a Wicked Problem ... 12

3.9 Reclassifying Aging Research as a Global Challenge 12
3.10 Conclusion 13
References 13

4 Investment in AgeTech 15
4.1 Introduction 15
4.2 The Case for AgeTech Investment 15
4.3 Global Investment Trends in AgeTech 16
4.4 Economic Potential of the Silver Economy 16
4.5 Examples of AgeTech Innovations 17
4.6 Challenges in AgeTech Investment 17
4.7 Strategies for Scaling AgeTech Investment 18
4.8 Case Studies in Successful AgeTech Investment 18
4.9 Conclusion 19
References 19

5 Public and Private Funding in Aging Research and Services: A Collaborative Imperative 21
5.1 Introduction 21
5.2 The Role of Public Funding 21
5.3 Private Sector Contributions 22
5.4 Collaborative Models of Funding 23
5.5 Strategies for Enhancing Funding Efficiency 23
5.6 Future Directions 24
5.7 Conclusion 24
References 24

6 Sustainable Funding for Aging Services: Securing the Future of Elder Care 27
6.1 Introduction 27
6.2 The Economic Impact of Aging Populations 27
6.3 Challenges in Current Funding Models 28
6.4 Current Models of Funding Aging Services 28
6.4.1 Publicly Funded Models 28
6.4.2 Private Funding Models 29
6.5 Strategies for Sustainable Funding 29
6.6 Case Studies 30
6.7 Future Directions 30
6.8 Conclusion 30
References 30

7 Why AgeTech Matters: Transforming the Future of Aging 33
7.1 Introduction 33
7.2 The Demographic Context of Aging 33
7.3 Healthcare Innovations in AgeTech 34
7.4 Enhancing Social Inclusion and Mental Health 35
7.5 Economic Benefits of AgeTech 35
7.6 Challenges and Barriers to Adoption 35
7.7 Policy and Advocacy for AgeTech 36
7.8 Case Studies: Success Stories in AgeTech 36
7.9 Conclusion 37
References 37

8 The Rise of AgeTech: Key Innovations, Assistive Devices, and Smart Home Solutions 39
8.1 Introduction 39
8.2 Key Innovations in Technology for Seniors 41
8.2.1 Wearable Health Devices 41
8.2.2 Telehealth Platforms 41
8.2.3 Cognitive Health Technologies 42
8.3 Assistive Devices for Seniors 43
8.3.1 Mobility Aids 43
8.3.2 Hearing and Vision Devices 43
8.4 Smart Home Solutions 44
8.4.1 IoT-Enabled Smart Homes 44
8.4.2 Fall Detection Systems 44
8.4.3 Home Healthcare Robotics 45
8.5 International Perspectives on AgeTech 45
8.6 Challenges and Ethical Considerations in AgeTech 46
8.6.1 Ethical Concerns Include 46
8.7 Conclusion 46
References 48

9 Digital Inclusion for Seniors: Bridging the Digital Divide and Designing User-Friendly Technologies 49
9.1 Introduction 49
9.2 The Digital Divide: Challenges Faced by Seniors 50
9.2.1 Access to Technology 50
9.2.2 Digital Literacy 50
9.2.3 Accessibility Challenges 50
9.2.4 Security and Privacy Concerns 51

9.3 The Importance of Digital Inclusion ... 51
9.3.1 Promoting Independence ... 51
9.3.2 Enhancing Social Connectivity ... 51
9.3.3 Facilitating Healthcare Access ... 52
9.3.4 Economic Participation ... 52
9.4 Bridging the Digital Divide: Global Strategies ... 52
9.4.1 Affordable Access Initiatives ... 52
9.4.2 Digital Literacy Training ... 53
9.4.3 Accessible Design Principles ... 53
9.4.4 Addressing Security Concerns ... 54
9.5 Designing User-Friendly Technologies ... 54
9.5.1 Human-Centered Design ... 54
9.5.2 Inclusive Product Testing ... 54
9.5.3 Adaptive Interfaces ... 55
9.6 Global Examples of Digital Inclusion ... 55
9.6.1 Canada ... 55
9.6.2 Japan ... 56
9.6.3 Europe ... 56
9.6.4 United States ... 56
9.7 Future Directions ... 57
9.8 Conclusion ... 57
References ... 58

10 AI and Robotics in Elder Care: Companions and Caregivers ... 59
10.1 Introduction ... 59
10.2 The Role of Artificial Intelligence in Elder Care ... 60
10.2.1 AI-Powered Health Monitoring ... 60
10.2.2 Virtual Healthcare Assistants ... 61
10.2.3 Cognitive Health Support ... 62
10.3 The Role of Robotics in Elder Care ... 62
10.3.1 Robotic Caregivers ... 62
10.3.2 Companion Robots ... 63
10.3.3 Service Robots ... 63
10.4 Companions and Caregivers: A Dual Role ... 64
10.4.1 Companionship for Social Well-Being ... 64
10.4.2 Enhanced Caregiving Through Automation ... 64
10.5 International Perspectives ... 65
10.5.1 Canada ... 65
10.5.2 Japan ... 65
10.5.3 United States ... 66
10.5.4 Europe ... 66

10.6 Challenges and Ethical Considerations ... 66
10.6.1 Cost and Accessibility ... 67
10.6.2 Data Privacy and Security ... 67
10.6.3 Balancing Automation and Human Touch ... 67
10.7 Future Directions ... 67
10.8 Conclusion ... 68
Reference ... 69

11 Ethical Considerations in AgeTech: Privacy, Security, and Balancing Automation with Human Touch ... 71
11.1 Introduction ... 71
11.1.1 Privacy and Security in Health Data ... 72
11.1.2 The Importance of Health Data Privacy ... 72
11.1.3 Cybersecurity Challenges ... 72
11.1.4 Informed Consent ... 73
11.1.5 On Data Security, Institutional Trust, and the Fragility of AgeTech Adoption ... 73
11.1.6 Solutions ... 74
11.2 Balancing Automation with Human Touch ... 76
11.2.1 The Role of Automation in Elder Care ... 76
11.2.2 The Emotional Dimension of Care ... 77
11.2.3 Designing Human-Centered Automation ... 77
11.2.4 Solutions ... 77
11.3 Ethical Implications of Assistive Technologies ... 78
11.3.1 Autonomy Versus Control ... 78
11.3.2 Bias and Accessibility ... 78
11.3.3 Solutions ... 79
11.4 International Frameworks for Ethical AgeTech ... 79
11.4.1 Canada ... 79
11.4.2 Japan ... 80
11.4.3 European Union ... 80
11.4.4 United States ... 80
11.5 Future Directions and Recommendations ... 80
11.5.1 Strengthening Ethical Governance ... 81
11.5.2 Enhancing Accessibility ... 81
11.5.3 Designing for Empathy ... 82
11.6 Conclusion ... 82

12 Is Scientific Peer Review Killing Innovation in AgeTech Solutions for Older Adults? ... 85
12.1 Introduction ... 85
12.2 The Current Landscape of AgeTech ... 87

12.2.1 Robotic Companions ... 87
12.2.2 Smart Home Technologies ... 88
12.2.3 Telehealth Platforms ... 89
12.3 Barriers to AgeTech Innovation ... 89
12.3.1 Lengthy Validation Processes ... 90
12.3.2 Resource Constraints ... 90
12.3.3 Regulatory Hurdles ... 91
12.3.4 Ethical and Legal Complexities ... 91
12.4 The Island of Misfit Solutions ... 92
12.4.1 AI-Powered Cognitive Tools ... 92
12.4.2 Robotic Caregivers ... 93
12.5 Proposed Reforms for Accelerating Innovation ... 94
12.5.1 Adaptive Regulatory Frameworks ... 94
12.5.2 Alternative Research Methodologies ... 94
12.5.3 Collaborative Models ... 95
12.5.4 Streamlined Ethical Oversight ... 95
12.5.5 Flexible Evidence Standards ... 96
12.6 Balancing Innovation with Ethical Integrity ... 96
12.6.1 Privacy and Data Security ... 96
12.6.2 Human-Centered Design ... 97
12.7 Global Perspectives ... 97
12.7.1 Canada ... 98
12.7.2 Japan ... 98
12.7.3 United States ... 99
12.7.4 Europe ... 99
12.8 Conclusion ... 100
References ... 101

13 Conducting Research in AgeTech: Paradigms, Approaches, and Challenges ... 103
13.1 Introduction ... 103
13.2 Research Paradigms in AgeTech ... 105
13.2.1 Quantitative Research Paradigms ... 105
13.2.2 Qualitative Research Paradigms ... 106
13.2.3 Mixed-Methods Approaches ... 107
13.3 User-Centered Design in AgeTech Research ... 108
13.3.1 Principles of User-Centered Design ... 108
13.3.2 Participatory Research ... 109
13.3.3 Usability Testing ... 109

13.4 Challenges in AgeTech Research ... 110
13.4.1 Recruitment and Retention ... 110
13.4.2 Ethical Considerations ... 111
13.4.3 Technological Adaptation ... 112
13.5 Interdisciplinary Collaboration in AgeTech Research ... 113
13.5.1 Engineering and Design ... 113
13.5.2 Healthcare ... 113
13.5.3 Social Sciences ... 114
13.6 Real-World Evidence in AgeTech Research ... 114
13.6.1 Observational Studies ... 114
13.6.2 Wearable Data Collection ... 115
13.6.3 Implementation Science ... 115
13.7 Future Directions in AgeTech Research ... 116
13.7.1 AI-Driven Research Tools ... 116
13.7.2 Cross-Cultural Studies ... 116
13.7.3 Hybrid Research Models ... 116
13.8 Conclusion ... 117
References ... 118

14 Why I Got into the AgeTech Field: A Personal Journey ... 119
14.1 Introduction ... 119
14.2 The Encounter that Changed Everything ... 119
14.3 Exploring Prompting Technologies and AI ... 120
14.4 Building a Career in AgeTech ... 120
14.5 Becoming a Caregiver for My Parents ... 120
14.6 Facing My Challenges: Aging with a Disability ... 121
14.7 Lessons Learned Along the Way ... 121
14.7.1 Empathy is Essential ... 121
14.7.2 Collaboration Fuels Innovation ... 121
14.7.3 Technology Alone is not Enough ... 122
14.8 The Future of AgeTech ... 122
14.9 Conclusion ... 122

15 Should I Go into a Career in AgeTech? ... 123
15.1 Introduction ... 123
15.2 Why AgeTech Matters? ... 124
15.2.1 Demographic Trends ... 124
15.2.2 Economic Impact ... 125
15.2.3 Societal Benefits ... 125

15.3 Careers in AgeTech ... 126
15.3.1 Research Careers in AgeTech ... 126
15.3.2 Industry Careers in AgeTech ... 127
15.3.3 Policy Careers in AgeTech ... 128
15.3.4 Startups and Entrepreneurship in AgeTech ... 129
15.4 Skills Needed for a Career in AgeTech ... 131
15.5 Global Perspectives on AgeTech Careers ... 131
15.5.1 Canada ... 131
15.5.2 Japan ... 132
15.5.3 United States ... 132
15.5.4 Europe ... 132
15.6 Challenges in AgeTech Careers ... 133
15.7 Future Directions in AgeTech ... 133
15.8 Conclusion ... 134
References ... 135

16 The Future of AgeTech: Bridging Science Fiction and Reality ... 137
16.1 Introduction ... 137
16.2 The Science Fiction of Aging ... 138
16.3 Redefining Independence Through Robotics ... 139
16.3.1 Robotic Ecosystems ... 139
16.3.2 Robotic Companions with Emotions ... 140
16.4 AI-Driven Healthcare Revolution ... 140
16.4.1 Predictive Health Systems ... 140
16.4.2 Digital Twins for Personalized Care ... 141
16.5 Immersive Virtual Realities for Aging ... 141
16.5.1 Escape Aging Through Virtual Worlds ... 141
16.5.2 Rehabilitation in the Metaverse ... 142
16.6 Biotechnological Advancements and Longevity ... 142
16.6.1 Regenerative Medicine ... 142
16.6.2 Genetic Reprogramming ... 143
16.7 Consciousness Preservation and Digital Immortality ... 143
16.7.1 Mind Uploading ... 143
16.7.2 AI Integration ... 143
16.8 AgeTech in Space Exploration ... 144
16.8.1 Older Astronauts ... 144
16.8.2 Interstellar Retirement Communities ... 145
16.9 Ethical Considerations for the Future ... 145
16.10 Conclusion ... 146
Reference ... 147

17 Interdisciplinary Collaboration: Bridging Gaps to Advance AgeTech Innovation 149
17.1 Introduction 149
17.2 The Necessity of Cross-Disciplinary Efforts in AgeTech 150
17.2.1 Complexity of Aging Challenges 150
17.2.2 Role of Other Disciplines 151
17.3 Case Studies from My Career 152
17.3.1 Intelligent Assistive Technology and Systems Lab (IATSL) 152
17.3.2 Development of COACH 152
17.3.3 Ambient Assisted Living 153
17.4 Bridging Engineering and Occupational Therapy 153
17.4.1 Application of Engineering Principles to Occupational Therapy 153
17.4.2 Enhancing Rehabilitation Through Technology 154
17.5 Challenges in Interdisciplinary Collaboration 154
17.5.1 Communication Barriers 154
17.5.2 Differing Perspectives and Priorities 155
17.6 The Impact of Interdisciplinary Collaboration on AgeTech Innovation 156
17.6.1 Enhanced User-Centered Design 156
17.6.2 Improved Technology Adoption 156
17.7 Future Directions 157
17.7.1 Continued Interdisciplinary Approaches 157
17.7.2 Potential Areas of Growth 157
17.8 My Vision for the Future 157
17.8.1 Initiatives to Promote Collaboration 157
17.8.2 Personal Reflection: Emphasizing the Human Element 158
17.9 Conclusion 158
References 159

18 Global Perspectives: Cultural Considerations in AgeTech 161
18.1 Introduction 161
18.2 Cultural Influences on AgeTech Adoption 162
18.2.1 Perceptions of Aging and Older Person Care 162
18.2.2 Technological Literacy and Accessibility 162
18.2.3 Societal Attitudes Toward Technology 163
18.3 Case Studies from the Global South 163
18.3.1 AgeTech in China: Balancing Tradition and Modernity 163
18.3.2 AgeTech in India: Addressing the Digital Divide 164
18.3.3 AgeTech in Latin America: Social Inclusion and Accessibility 164

18.3.4 AgeTech in Africa: Innovative Solutions Amidst Challenges ... 164
18.4 Designing Culturally Appropriate AgeTech Solutions ... 165
18.4.1 User-Centered Design with Cultural Sensitivity ... 165
18.4.2 Overcoming Language and Literacy Barriers ... 165
18.4.3 Building Trust and Addressing Privacy Concerns ... 165
18.4.4 Economic Considerations ... 165
18.5 Impact of Cultural Differences on Specific AgeTech Domains ... 166
18.5.1 Telehealth and Telemedicine ... 166
18.5.2 Robotics and AI in Older Person Care ... 166
18.5.3 Wearable Technologies ... 166
18.6 Strategies for Successful Implementation of AgeTech in Diverse Cultural Contexts ... 167
18.6.1 Education and Training ... 167
18.6.2 Policy and Government Support ... 167
18.6.3 Collaboration with Local Organizations ... 167
18.7 Conclusion ... 167
References ... 168

19 Education and Training for the Next Generation of AgeTech Professionals ... 171
19.1 Introduction ... 171
19.2 The Need for Specialized Education in AgeTech ... 172
19.2.1 Demographic Changes and Aging Population ... 172
19.2.2 Technological Advancements and Opportunities ... 172
19.2.3 Skill Requirements in AgeTech ... 172
19.3 Models of Education and Training in AgeTech ... 173
19.3.1 Theoretical Learning ... 173
19.3.2 Experiential Learning ... 174
19.4 Examples of Training Programs in AgeTech ... 174
19.4.1 EPIC in AGE-WELL ... 174
19.4.2 Other Notable Programs ... 175
19.5 Strategies for Preparing Future AgeTech Professionals ... 176
19.5.1 Interdisciplinary Education ... 176
19.5.2 Collaboration Between Academia and Industry ... 176
19.5.3 Incorporating User-Centered Design and Ethics ... 177
19.5.4 Ethical Considerations ... 177
19.5.5 User-Centered Design Projects ... 177
19.6 Challenges and Solutions ... 178

19.6.1 Keeping Curriculum up to Date with Rapid Tech Changes 178
19.6.2 Ensuring Accessibility and Inclusivity in Education 178
19.6.3 Overcoming Barriers to Experiential Learning 178
19.7 Conclusion 179
References 179

20 Will We Ever See Anything Novel Again in AgeTech? 181
20.1 Introduction 181
20.2 The State of AgeTech: An Era of Incrementalism 182
20.2.1 A Proliferation of Redundancy 182
20.2.2 The Comfort of the Known 182
20.3 The Innovator's Dilemma in AgeTech 183
20.3.1 AgeTech's Entrapment in Sustaining Innovations 183
20.3.2 The Risk of Disruption from Outside 183
20.4 The Need for Novelty: Beyond Incremental Improvements 184
20.4.1 Reimagining Problems and Solutions 184
20.4.2 Embracing Failure as a Pathway to Innovation 184
20.5 Trusting the Experts: Valuing Professional Insight 185
20.5.1 The Limits of User-Centered Design 185
20.5.2 The Role of Visionary Leadership 185
20.6 Provoking Change: A Call to Action for the AgeTech Field 185
20.6.1 Fostering an Innovation Ecosystem 185
20.6.2 Re-Evaluating Success Metrics 186
20.6.3 Educating the Next Generation 186
20.7 Conclusion 186
References 187

Conclusion: Breaking the Cycle of Incrementalism in AgeTech 189

Recommended Readings for Each Essay 191

About the Author

Alex Mihailidis Ph.D., P.Eng. FCAHS, is a Professor at the University of Toronto and the Inaugural Scientific Director of AGE-WELL, Canada's national network on technology and aging. He also serves as the Associate Vice-President for International Partnerships at U of T. With over 20 years of experience, Alex's research bridges health sciences, engineering, and social policy to create technologies that support older adults and caregivers. He has published extensively and leads national and international initiatives focused on aging with dignity.

His work is driven by the belief that technology should enhance—not replace—human care. He is a Fellow of the Canadian Academy of Health Sciences and RESNA, and in 2022 was named one of the United Nations' Healthy Ageing 50 global leaders. In 2025, he was named a Chevalier of the Ordre des Palmes Académiques by the French Republic in recognition of his contributions to international academic collaboration and education.

Beyond his academic and leadership roles, Alex is a committed mentor and collaborator. The ideas in this book are informed by years of dialogue with students, colleagues, older adults, and his own experiences as a caregiver. He lives in Toronto with his wife and two children, who continue to inspire his work and remind him why building a more compassionate future for aging matters.

1 Why Did I Write These Essays? More Importantly, Why Should You Read Them?

1.1 Setting the Scene: A Personal Reflection

Writing these essays on aging and AgeTech was not born from a casual interest but a strong conviction. As a society, we face one of the most transformative demographic shifts in human history: the aging population. According to the United Nations, by 2050, one in six people in the world will be over the age of 65. The implications are profound, not just for healthcare and economics but for how we perceive life itself.

So why did I write these essays? Aging is not just a biological inevitability; it's an already-existing frontier of innovation, empathy, and social transformation. These essays are my intellectual rebellion against complacency that I have seen growing in the aging and AgeTech worlds, a manifesto for those who dare to envision a future where aging is not a decline but a renaissance. AgeTech is a tool that can help us achieve what we all want as we age.

But why make these essays essential reading? Because they aren't just about my journey or my convictions. They are about you, your family, and our world. They explore questions we often avoid: How do we ensure dignity in aging? What role should technology play in preserving humanity as it enhances functionality? And most provocatively, how do we prevent AgeTech from becoming a dystopian commodification of old age? Each essay is a piece of this puzzle, or at least I hope they are.

The following is an outline of these essays and their contents. They do not need to be read in order or out of a sense of duty to the field. They can be read based on your interests, whether you are a researcher, in industry, a policymaker, an entrepreneur, a student, or an older adult. Most importantly, each essay is short and not loaded with specific jargon, scientific references, or theories. They are based on my thoughts, experiences, and what I have learned over my career. Mainly, though, they are based on important lessons gained from working with those we are developing AgeTech for older people and their

A. Mihailidis, *The Age of AgeTech*, Synthesis Lectures on Technology and Health,
https://doi.org/10.1007/978-3-032-04886-8_1

caregivers. This collective work is truly a culmination of things that I have written over the years but have never attempted to publish. They are a collection of hand-written notes, scraps of paper in my journals, and the back-of-the-napkin ideas.

1.2 Overview of Essays

The first essay (Essay 2) is a wake-up call, a mirror held up to a world that often ignores the unavoidable truth of aging. I wrote it not to offer comfortable platitudes but to dismantle societal complacency and force a reckoning with the profound implications of a rapidly aging global population. It tries to boldly challenge the reader to confront the realities of demographic shifts, the strain on healthcare and economies, and the cultural blind spots that undervalue older adults.

The essay declares that aging is not an issue for "someone else," it is everyone's reality, an unrelenting force that shapes economies, policies, and human lives. If you ignore it, you are ignoring the future itself. Addressing the challenges and opportunities, such as the booming silver economy and the potential for intergenerational strength, provokes a radical rethinking of how societies should embrace aging as a transformative force, not a looming crisis.

Essay 3 attempts to explore questions like why do we invest trillions in prolonging youth but mere scraps into understanding aging? In "Addressing the Imbalance," I argue that the disparity in funding for aging research reflects a fiscal oversight and a cultural failure to value the latter stages of life. This imbalance isn't just a research gap—it's a moral chasm. Without proper funding, we fail to unlock the secrets of aging that could lead to breakthroughs benefiting all generations. Shouldn't understanding longevity and its complexities be as alluring to humanity as landing on Mars?

Essay 4 attempts to argue for an increase in investment in AgeTech. If aging is the problem, AgeTech is the solution, or, more accurately, at least part of it. In "Investment in AgeTech," I explore why AgeTech startups still struggle to attract funding despite the colossal opportunity they represent. This is not just about venture capital; it's about recalibrating our collective vision. If you think AgeTech is merely about walking aids or better wheelchairs, you are missing the forest for the trees. Imagine smart homes that adapt to their residents, wearables that predict health crises, and digital companions that combat loneliness. This essay challenges us to think like a futurist: Why are we NOT betting on AgeTech?

Essay 5, "Public and Private Funding in Aging Research," dives into the dependency paradox. Governments alone cannot shoulder the financial burden of an aging population. Conversely, leaving it entirely to the private sector risks turning aging into a profit-driven endeavor devoid of compassion. I advocate for collaborative imperatives—partnerships where the public good aligns with private innovation. Specifically, are we ready to transcend political and economic silos to tackle aging collectively?

The next essay examines one of the most glaring weaknesses in our current systems: the unsustainable funding of aging services. Elder care is at risk of becoming a luxury, accessible only to the wealthy. But what if elder care was treated as an infrastructure project? As we build roads and bridges, investing in aging services creates societal dividends: healthier populations, reduced healthcare costs, and empowered communities. This essay attempts to ask: Is it time for a New Deal for aging?

Essays 7 through 10 map the landscape of AgeTech, from innovation to ethics. "Why AgeTech Matters" is a call for integrating technology into aging to transform care and entire lives. "The Rise of AgeTech" showcases key innovations, proving that AgeTech is no longer speculative—it is here and evolving rapidly.

"Digital Inclusion for Seniors" confronts the digital divide. It is easy to romanticize the potential of AgeTech, but what good is a smart device if a senior does not know how to use it? This essay outlines practical strategies to ensure no one is left behind in the tech revolution.

"AI and Robotics in Elder Care" takes a provocative turn, asking whether we are ready for robots to be caregivers and companions. Is AI the antidote to loneliness, or does it threaten to erode human connections? The answers aren't simple, but they are vital.

Essay 11, "Ethical Considerations in AgeTech," explores the toughest questions: How do we balance automation with the human touch? How do we safeguard privacy in an age of intrusive technologies? I hope this essay encourages you to question your assumptions and consider the fine line between progress and exploitation.

In my opinion, Essay 12 "Is Scientific Peer Review Killing Innovation in AgeTech?" may be my most controversial piece. While peer review ensures rigor, it often stifles radical and potentially innovative ideas. What if the most transformative AgeTech solutions languish in unpublished papers because they challenge the status quo? This essay doesn't just critique; it proposes solutions, urging the scientific community to embrace a culture of bold experimentation. It forces the scientific community, including myself, to rethink how we assess and measure the success of our field.

The remaining Essays 13, 14, 15, 16, 17, 18, 19 and 20, explore broader questions. "Conducting Research in AgeTech" emphasizes the interdisciplinary nature of this field, while "Should I Go Into a Career in AgeTech?" serves as both a guide and a rallying cry for aspiring professionals.

In "Why I Got Into the AgeTech Field," I share my journey—what drew me to this intersection of technology and humanity. It's a story of curiosity, heartbreak, and hope.

"The Future of AgeTech" blends science fiction with reality, imagining a world where aging is not feared but embraced. In "Interdisciplinary Collaboration," I argue that the future of AgeTech depends on breaking silos. No single discipline can solve the complexities of aging; only collaboration can address them effectively.

"Global Perspectives" offers a cultural lens, examining how different societies approach aging. It reminds us that AgeTech is not one size fits all; it must adapt to diverse values and traditions.

Finally, "Education and Training" and "Will We Ever See Anything Novel Again in AgeTech?" focus on the future. How do we prepare the next generation to tackle these opportunities and challenges? And how do we ensure that AgeTech remains a hotbed of innovation rather than a stagnant field?

At the conclusion of this collection, you will find a dedicated section offering carefully curated additional readings for each essay. These resources, ranging from landmark studies to accessible articles, are designed to deepen your understanding, spark further reflection, and connect you with the broader conversations happening across disciplines in aging, technology, design, ethics, and policy. Whether you are a curious reader, a student, or a professional in the field, this reading list serves as a springboard for continued exploration.

1.3 Why Should You Care?

Why should you read these essays? Perhaps because you have some time on your hands. But hopefully it's because you know that aging affects everyone. Whether you are 25 or 85, aging is the universal experience we all share. These essays aren't just about seniors, they are about reimagining society. I hope that they challenge you to think differently about technology, equity, and what it means to live a fulfilling life at any age.

This book caters to a diverse audience interested in the future of aging and how technology influences it. It is aimed at researchers and academics seeking a wider context for their work, entrepreneurs and policymakers trying to understand the direction of the field, and students exploring careers in AgeTech. It is also for older adults, caregivers, and advocates who wish to see their experiences reflected in thoughtful, forward-looking conversations. Most importantly, it is for anyone who cares about how society values and prepares for the process of growing older.

The stakes could not be higher. Without action, we risk a future in which aging is synonymous with inequality, neglect, and despair. However, with vision, investment, and collaboration, we can create a world where aging is synonymous with wisdom, innovation, and dignity.

2 If You're Not Aging, You're Dead

Aging is a universal truth. From the moment we are born, our bodies, minds, and lives follow a path of growth, change, and eventual decline. The phrase "*If you're not aging, you're dead*" (a phrase I often share in my presentations and discussions with my research team) might seem blunt, but it highlights a fundamental reality: aging is the cost of living. Although aging is unavoidable, how it influences our world, biologically, socially, and demographically, is quite complex. Investigating what aging signifies, especially in a world where older populations are expanding rapidly, helps us better understand the potential impacts and opportunities that come with growing older.

2.1 The Science Behind Aging

Aging starts at the cellular level, where processes that sustain life gradually lose efficiency over time. One of the major culprits is cellular senescence, which is the point at which our cells stop dividing, leading to tissue degeneration [5]. Scientists have also identified oxidative stress, damage caused by reactive oxygen species (ROS), as a significant contributor [6]. Think of it as rust forming on a piece of metal over time; the same happens to our cells.

Interestingly, research on telomeres, the protective caps on our chromosomes, has given us insights into how our cells "*count down*" their lifespan [1]. As telomeres shorten, cells lose their ability to divide effectively, a hallmark of aging. Additionally, mitochondrial dysfunction, the decline of the energy centers in our cells, plays a role in fatigue and other age-related changes. Understanding these processes sheds light on why we age and helps guide research into how we might age more healthily.

A. Mihailidis, *The Age of AgeTech*, Synthesis Lectures on Technology and Health,
https://doi.org/10.1007/978-3-032-04886-8_2

2.2 How Aging Feels?

Beyond biology, and perhaps more importantly, aging is a deeply personal experience. For some, it brings wisdom, self-assurance, and a sense of freedom. For others, it's accompanied by fear of irrelevance, physical and cognitive decline, or isolation. Erik Erikson, a renowned psychologist, described late adulthood as a stage of reflection where people grapple with a sense of integrity versus despair [4]. Aging can be bittersweet, balancing the joy of accumulated experiences with the recognition of life's finality—death.

2.3 A Global Phenomenon

Aging populations are one of the most significant demographic shifts of our time. According to the United Nations [7], the number of people aged 60 and older is expected to double by 2050, reaching a staggering 2.1 billion [7]. That is not just an abstract number; it represents millions of families, communities, and economies adapting to a new reality.

In Japan, for instance, nearly 30% of the population is over 65. This aging trend has led to labor shortages and a booming market for elder care robots—innovations that feel like science fiction but are very real. Similarly, European countries such as Germany and Spain (where I currently sit while writing this document) face opportunities and challenges with sustaining their healthcare systems and pension funds as their populations grow older and birth rates decline [7].

Other nations like China and India are also grappling with rapid aging, but their situations differ. In China, the one-child policy has compounded the aging issue, leaving fewer younger people to care for a burgeoning elderly population. India, while still younger on average, will face similar opportunities and challenges in the coming decades as life expectancy increases.

2.4 Aging in Canada

The aging trend in Canada is just as striking. Nearly 18% of Canadians are 65 or older as of 2023, a figure expected to rise to 25% by 2036. At 82 years, life expectancy in Canada is among the highest in the world, both a testament to advances in healthcare and a challenge for policymakers.

While urban areas like Toronto and Vancouver attract younger populations, rural communities are aging faster, creating regional imbalances. Indigenous communities, on the other hand, present a unique demographic story. With higher birth rates and younger populations overall, they face different opportunities and challenges, including exploring disparities in life expectancy and access to care.

2.5 Why Aging Matters to Everyone?

The implications of aging populations reach far beyond older people themselves. Economically, a shrinking workforce combined with increased healthcare demands strains public resources. Consider pensions: They are essentially a social contract in which today's workers fund the retirement of older generations. The system faces financial stress as the ratio of workers to retirees shrinks.

Healthcare costs are another major issue. Older adults require more frequent and intensive medical care, which puts pressure on healthcare systems worldwide [3]. With its universal healthcare model, Canada faces significant opportunities and challenges in funding long-term care and improving infrastructure for an aging population.

But it's not just about costs and systems—culture and community are equally, if not more, important. Older adults play vital roles in family life, often as caregivers for grandchildren or repositories of cultural knowledge [2]. Societies that embrace intergenerational relationships tend to thrive, creating spaces where everyone feels valued and included regardless of age.

2.6 The Need for Innovation and Policy Change

Governments and institutions need to act decisively to explore the opportunities and challenges of aging populations. In Canada, reforms to pension systems and investments in elder care are essential. Policies encouraging older adults to stay in the workforce longer, such as flexible retirement options, can also help balance the labor market.

Urban planning must also adapt. Age-friendly cities, including educational institutions like I belong to, prioritizing accessibility, public transportation, and green spaces improve life for everyone—not just older adults. Globally, innovations like Japan's elder care robots or Germany's focus on intergenerational housing provide models for other countries to learn from.

2.7 Opportunities Amid Challenges

While aging populations have opportunities and challenges, they also offer unique opportunities. For one, the "silver economy", industries catering to older adults, is booming. From healthcare technology to travel services, businesses are finding innovative ways to meet the needs of this demographic. Telemedicine, for example, is revolutionizing healthcare access for people with mobility issues or chronic conditions.

Societally, older adults contribute immensely through volunteering, mentorship, and civic engagement. Retirees often have the time and experience to support their communities in ways that working-age individuals cannot. Recognizing and leveraging this potential is key to creating vibrant, inclusive societies.

2.8 A Shared Responsibility

At its heart, the issue of aging is one of shared responsibility. Every individual, community, and government has a role to play in exploring the opportunities and challenges of this demographic shift. This is not just about older people, it's about creating a world where people of all ages can thrive.

Aging is universal. It connects us to our past through the wisdom of older generations and to our future through the legacy we leave behind. Understanding and embracing this reality can build stronger, more compassionate societies celebrating life at every stage.

References

1. Blackburn, E., & Epel, E. (2017). *The telomere effect: A revolutionary approach to living younger, healthier, longer*: Hachette UK.
2. Carstensen, L. L., Isaacowitz, D. M., & Charles, S. T. (1999). Taking time seriously: A theory of socioemotional selectivity. *American Psychologist, 54*(3), 165–181. https://doi.org/10.1037/0003-066X.54.3.165
3. Centers for Medicare & Medicaid Services. (2024). National Health Expenditure Data. Retrieved from https://www.cms.gov/data-research/statistics-trends-and-reports/national-health-expenditure-data
4. Erikson, E. H. (1998). *The life cycle completed (extended version)*: WW Norton & Company.
5. Lee, S. H., Lee, J. H., Lee, H. Y., & Min, K. J. (2019). Sirtuin signaling in cellular senescence and aging. *BMB Rep, 52*(1), 24–34. https://doi.org/10.5483/BMBRep.2019.52.1.290
6. Miwa, S., Kashyap, S., Chini, E., & von Zglinicki, T. (2022). Mitochondrial dysfunction in cell senescence and aging. *The Journal of clinical investigation, 132*(13). https://doi.org/10.1172/JCI158447
7. United Nations. (2024). World Population Prospects 2024. Retrieved from https://www.un.org/development/desa/pd/world-population-prospects-2024

3 Addressing the Imbalance: The Urgent Need for Increased Funding in Aging Research

3.1 Introduction

The twenty-first century is witnessing an unprecedented demographic shift characterized by rapidly aging populations. According to the World Health Organization (WHO), the number of individuals aged 60 and older will increase from 1 billion in 2020 to 2.1 billion by 2050 [6]. This transformation has significant societal implications, influencing labor markets, healthcare systems, and social support structures. The growing prevalence of aging-related conditions such as dementia, cardiovascular disease, and arthritis places an enormous strain on healthcare systems, exacerbating costs and workforce shortages.

Despite the magnitude of these opportunities and challenges, research on aging receives significantly less funding. In 2020, the National Institute on Aging (NIA) in the United States received $3.54 billion, far less than the $6.56 billion allocated for cancer research by the National Institutes of Health (NIH). This essay examines the funding disparity in aging research, its societal and economic consequences, and the need for reclassifying aging as a global research priority.

3.2 The Demographic Shift Toward an Aging Population

The aging population results from declining fertility rates and increasing life expectancy. Advances in healthcare, sanitation, and nutrition have contributed to lower mortality rates, particularly among younger age groups, allowing more individuals to reach old age [2]. By 2050, one in six people worldwide will be over 65, rising to one in four in Europe and North America [5].

While developed countries experience the highest absolute numbers of older adults, developing nations are witnessing the fastest growth in aging populations. For instance,

A. Mihailidis, *The Age of AgeTech*, Synthesis Lectures on Technology and Health,
https://doi.org/10.1007/978-3-032-04886-8_3

countries like China, Brazil, and India face unique opportunities and challenges as they transition from predominantly young to aging societies. These changes demand robust social and healthcare systems to accommodate the needs of older citizens.

3.3 The Impact on Healthcare Systems

The healthcare utilization of older adults is significantly higher than that of younger populations. In the United States, individuals aged 65 and older accounted for 36% of total healthcare expenditures in 2019, despite constituting only 16% of the population [1]. This trend is mirrored globally, with aging populations contributing to chronic disease prevalence and long-term care demands.

Aging populations exacerbate shortages of healthcare workers. The WHO estimates that by 2030, there will be an 18 million-person shortfall in healthcare professionals, particularly in low- and middle-income countries [6]. This gap emphasizes the need for specialized training and workforce planning to support aging populations effectively [3].

3.4 Funding Disparities: Aging Research Versus Other Areas

Despite the profound societal and economic impact of aging, research into aging receives significantly less funding compared to diseases such as cancer, cardiovascular disease, and diabetes (see Table 3.1). This discrepancy arises partly from the perception of aging as a "natural" process rather than a modifiable or treatable condition [4]. However, given that aging is the single largest risk factor for most chronic diseases, greater investment in aging research could yield broad health benefits by delaying the onset of multiple diseases simultaneously.

Table 3.1 Comparison of research funding across major disease areas versus aging

Research area	U.S. NIH annual funding (2022)	Global estimated annual funding	% of total NIH budget (%)
Cancer	$7.2 billion	~$80 billion (global)	17
Cardiovascular disease	$2.9 billion	~$50 billion (global)	7
Diabetes	$1.2 billion	~$30 billion (global)	3
Alzheimer's disease	$3.5 billion	~$40 billion (global)	8
Aging research (general)	$400 million	~$2 billion (global)	1

Aging research consistently lags behind in funding, even though age-related diseases such as cancer, cardiovascular disease, and neurodegenerative conditions impose substantial costs on healthcare systems.

3.5 Healthcare Expenditures: Aging Versus Disease-Specific Costs

Aging-related healthcare costs frequently exceed those of specific diseases due to the accumulation of chronic conditions in older populations:

- **United States**: Medicare spending on individuals aged 65+ accounted for 55% of total healthcare expenditures in 2021.
- **Japan**: Older adults (65+) represent 60% of healthcare spending, highlighting the financial burden of an aging population.
- **Europe**: Age-related diseases contribute to over 70% of healthcare costs, with multimorbidity driving higher per-capita expenditures.

3.5.1 Potential Economic and Health Benefits of Increased Aging Research Funding

Increasing funding for aging research could have widespread benefits, including

- Delaying multiple age-related diseases simultaneously rather than treating them individually.
- Reducing long-term healthcare costs by preventing or slowing the progression of chronic diseases.
- Enhancing workforce productivity by extending healthy lifespan and reducing disability rates.
- Developing novel longevity therapies that could mitigate frailty, cognitive decline, and other age-related impairments.

3.6 Barriers to Increased Funding for Aging Research

Aging is often seen as inevitable, reducing its perceived urgency as a research priority. This perception contrasts with conditions like cancer, which are viewed as treatable and preventable, garnering greater public and philanthropic support.

The multifactorial nature of aging complicates research efforts. Aging involves genetic, environmental, and lifestyle factors, requiring interdisciplinary approaches that may not yield immediate results.

Interventions targeting aging processes raise moral questions, including concerns about overpopulation and resource allocation. These debates can hinder funding and policy consensus.

3.7 The Case for Increased Investment in Aging Research

Many chronic diseases share underlying biological mechanisms associated with aging. Research into these mechanisms could lead to therapies that explore multiple conditions simultaneously, reducing healthcare costs and improving quality of life.

Investing in aging research could yield substantial financial returns. Delaying the onset of aging-related diseases by just a few years could save trillions in healthcare expenditures. A study in Health Affairs estimated that delayed aging could save $7.1 trillion over 50 years in the United States alone.

Finally, improving healthspan, the period of life spent in good health, could enable older adults to remain active and productive members of society, reducing dependency ratios and boosting economic growth.

3.8 Aging as a Wicked Problem

Aging exemplifies a "wicked problem" characterized by complexity, uncertainty, and interdependent factors. Solutions in one domain (e.g., healthcare) often influence other areas (e.g., social security systems), requiring holistic approaches.

Addressing aging requires collaboration across disciplines and sectors, from biology and healthcare to economics and urban planning. This interdisciplinary focus underscores the need for coordinated investment and policymaking.

3.9 Reclassifying Aging Research as a Global Challenge

Designating aging as a grand challenge could attract government, industry, and philanthropy funding. Initiatives like the WHO's Healthy Aging strategy emphasize the importance of global collaboration.

Raising public awareness of aging issues can increase support for research initiatives. Advocacy groups and campaigns can play a vital role in shifting public perceptions and influencing funding priorities.

3.10 Conclusion

The aging population represents one of the most significant challenges of the twenty-first century, yet aging research remains underfunded. Addressing this imbalance is essential to reducing healthcare costs, improving quality of life, and fostering innovation. Recognizing aging as a global research priority can mobilize resources and collaborations to tackle this issue effectively. Increased investment in aging research is not just a necessity, it is an opportunity to redefine health and longevity for future generations.

References

1. Centers for Medicare & Medicaid Services. (2024). National Health Expenditure Data. Retrieved from https://www.cms.gov/data-research/statistics-trends-and-reports/national-health-expenditure-data
2. Christensen, K., Doblhammer, G., Rau, R., & Vaupel, J. W. (2009). Ageing populations: the challenges ahead. *The Lancet, 374*(9696), 1196-1208. https://doi.org/10.1016/S0140-6736(09)61460-4
3. Fang, E. F., Xie, C., Schenkel, J. A., Wu, C., Long, Q., Cui, H.,... Woo, J. (2020). A research agenda for ageing in China in the 21st century (2nd edition): Focusing on basic and translational research, long-term care, policy and social networks. *Ageing Research Reviews, 64*, 101174. https://doi.org/10.1016/j.arr.2020.101174
4. López-Otín, C., Blasco, M. A., Partridge, L., Serrano, M., & Kroemer, G. (2013). The Hallmarks of Aging. *Cell, 153*(6), 1194-1217. https://doi.org/10.1016/j.cell.2013.05.039
5. United Nations. (2024). World Population Prospects 2024. Retrieved from https://www.un.org/development/desa/pd/world-population-prospects-2024
6. WHO. (2015). *World report on aging and health.* Retrieved from Geneva: https://www.who.int/publications/i/item/9789241565042/

4 Investment in AgeTech

4.1 Introduction

AgeTech, short for aging technology, encompasses innovations designed to improve the lives of older adults while addressing the opportunities and challenges associated with aging. These technologies span various domains, including healthcare, accessibility, and financial tools, offering solutions that promote autonomy, reduce caregiving burden, and enhance healthcare efficiency. As the global population ages at an unprecedented pace, AgeTech has emerged as a critical sector with immense potential for social and economic impact. However, despite its promise, investment in AgeTech remains significantly lower compared to other technology and healthcare sectors [1].

This essay explores the current state of investment in AgeTech, emphasizing its economic potential, existing barriers, and strategies for attracting more funding. It also examines global trends and case studies highlighting AgeTech's transformative role in the twenty-first century.

4.2 The Case for AgeTech Investment

The United Nations predicts that by 2050, over 2 billion people will be 60 years or older, making up nearly one-fifth of the global population [2]. This demographic shift profoundly impacts economies, healthcare systems, and social structures. Age-related healthcare expenditures already account for up to 40% of total healthcare spending in many developed nations [3]. Despite these pressing needs, AgeTech receives far less attention and funding than sectors like fintech or edtech [4]. Increasing targeted investment in AgeTech is essential to meet the growing demands of aging populations and capitalize on the opportunities presented by the burgeoning "silver economy" [5].

A. Mihailidis, *The Age of AgeTech*, Synthesis Lectures on Technology and Health,
https://doi.org/10.1007/978-3-032-04886-8_4

AgeTech's significance stems from its ability to address critical aspects of aging. From healthcare innovations that extend life expectancy to technologies that promote social connectivity, AgeTech solutions are positioned to improve the quality of life for older adults globally. The financial implications of underfunding this sector are considerable, as governments and businesses face escalating costs related to elder care and reduced productivity due to workforce imbalances. By addressing these gaps, AgeTech has the potential to transform societal attitudes toward aging and create more inclusive, resilient communities.

4.3 Global Investment Trends in AgeTech

Although venture capital investment in AgeTech has grown in recent years, it remains modest compared to other sectors. In 2020, global venture funding in AgeTech reached $2.5 billion, a notable increase from $1 billion in 2017. This growth demonstrates the increasing recognition of AgeTech's potential, yet the figures remain small compared to investments in areas such as artificial intelligence or e-commerce. The disparity underscores the need for greater advocacy and strategic funding to position AgeTech as a priority sector [6].

Regionally, the United States leads in AgeTech investment, with a strong focus on startups specializing in eldercare robotics, telemedicine, and financial services. Europe has also made significant progress through public–private partnerships, particularly in the Nordic countries, which are known for their innovative approaches to elder care. In Asia, countries like Japan and South Korea have emerged as leaders in AgeTech innovation, driven by their rapidly aging populations and advanced technological infrastructure [1]. These regional variations highlight the importance of tailoring AgeTech investments to local demographic and economic conditions [7].

4.4 Economic Potential of the Silver Economy

The silver economy, encompassing all economic activities targeting older adults, is projected to grow to $15 trillion globally by 2030. This expansive market spans multiple sectors, including healthcare, travel, housing, and financial services, all tailored to the needs of an aging population. The economic potential of this market is immense, presenting opportunities for businesses and governments to drive growth while addressing critical societal needs.

Key growth areas within the silver economy include healthcare and assistive technologies. Innovations such as wearable health monitors, telehealth platforms, and remote care solutions are leading the charge in improving healthcare access and outcomes for older adults. Mobility and accessibility solutions, including smart homes, advanced mobility aids, and public transport options designed for seniors, are also advancing rapidly. Additionally, financial technologies are increasingly enabling older adults to manage their finances, investments, and fraud prevention, contributing to their financial independence and security [8].

4.5 Examples of AgeTech Innovations

Several groundbreaking innovations exemplify the transformative potential of AgeTech. ElliQ by Intuition Robotics is an AI-powered companion robot designed to combat loneliness among older adults by facilitating social interaction and engagement. GreatCall, acquired by Best Buy for $800 million in 2018, provides senior-friendly mobile devices and health-tracking tools, showcasing the commercial viability of AgeTech products. Additionally, remote patient monitoring devices like AliveCor and Omron HeartGuide are being used to manage chronic conditions such as hypertension and arrhythmia, offering a glimpse into the future of personalized healthcare for aging populations.

These examples illustrate how AgeTech solutions address specific challenges faced by older adults, such as social isolation, declining health, and limited mobility. They also highlight the sector's potential to generate significant economic returns by meeting the needs of a rapidly growing demographic. By scaling these innovations, stakeholders can create a more inclusive and supportive environment for older adults, ensuring that technological advancements benefit all segments of society.

4.6 Challenges in AgeTech Investment

Despite its promise, several challenges hinder the growth of AgeTech investment. One of the most significant obstacles is the regulatory landscape. Many AgeTech innovations fall under stringent healthcare regulations, which can delay market entry and deter potential investors. This continues to be a significant issue in Canada.

Another major challenge is the digital divide among older adults. While AgeTech solutions hold immense potential, their adoption is often limited by the lack of digital literacy among seniors. In the U.S., only 61% of individuals aged 65 and older own a smartphone, compared to 95% of those aged 18–29 [9]. This disparity in digital access and skills reduces the effectiveness and reach of many AgeTech products.

4.7 Strategies for Scaling AgeTech Investment

To overcome these challenges and unlock the full potential of AgeTech, targeted strategies are essential. One effective approach is to promote public–private partnerships. Collaborative efforts between governments and private companies can foster innovation and drive investment in AgeTech. For example, the Japanese government has actively supported robotics startups developing eldercare solutions, such as exoskeletons and companion robots. Similarly, the EU's Horizon 2020, and now the Horizon Europe Program has allocated significant funding to aging-related technologies and services, demonstrating the potential of coordinated investment.

De-risking investments is another critical strategy. Governments can play a pivotal role in reducing financial risks for investors by offering tax credits, subsidies, and grants for AgeTech research and development. Programs like the U.S. Small Business Innovation Research (SBIR) grants have successfully supported early-stage startups, enabling them to bring innovative products to market.

Expanding digital literacy among older adults is equally important. Initiatives such as Singapore's "Seniors Go Digital" program, which provides free workshops and training for older adults, can bridge the digital divide and facilitate the adoption of AgeTech solutions. By improving digital skills, these programs empower seniors to engage with technology, enhancing their quality of life and creating new opportunities for businesses.

Fostering impact investment is another promising avenue. Impact investors, who prioritize social benefits alongside financial returns, are well suited to support AgeTech initiatives. The Rockefeller Foundation's investment in scalable aging-related solutions is a testament to the potential of this approach. By aligning financial incentives with societal needs, impact investing can drive sustainable growth in the AgeTech sector.

4.8 Case Studies in Successful AgeTech Investment

Several case studies illustrate the success and potential of AgeTech investment. Google's Calico Labs, founded in 2013, focuses on combating aging and related diseases through advanced biological research. Backed by significant investment from Alphabet, Calico Labs exemplifies the long-term commitment required to drive innovation in AgeTech [1].

In Japan, companies like Cyberdyne have leveraged government support to develop groundbreaking robotics solutions for eldercare. Cyberdyne's exoskeletons, designed to improve mobility for seniors, highlight the transformative potential of AgeTech innovations [6]. These examples underscore the importance of strategic investment and collaboration in advancing the AgeTech sector.

4.9 Conclusion

Investment in AgeTech represents both a necessity and an opportunity. As aging populations reshape global demographics, the demand for innovative solutions will only increase. By addressing barriers such as regulatory complexities and digital literacy gaps, stakeholders can unlock AgeTech's full potential. Public–private collaborations, impact investing, and targeted policies will be crucial to fostering a robust AgeTech ecosystem. The projected growth of the silver economy highlights vast opportunities for investors. Aligning financial incentives with social impact can create a win–win scenario, ensuring a healthier and more inclusive future for older generations.

References

1. Sixsmith, A. (2021). AgeTech: Technology-based solutions for aging societies. *Promoting the health of older adults: the Canadian experience, 135.*
2. United Nations. (2024). World Population Prospects 2024. Retrieved from https://www.un.org/development/desa/pd/world-population-prospects-2024
3. WHO. (2015). *World report on aging and health.* Retrieved from Geneva: https://www.who.int/publications/i/item/9789241565042/
4. Bhattacharya-Craven, A., Jackson, R., & Schanz, K.-U. (2022). Financial Wellbeing: Is it the key to reinventing life insurance? *Zurich: The Geneva Association.*
5. Schafer, D., & Rutley, J. (2021). *Aging Populations: The Role Of Venture Capital In Supporting The Industry.* Retrieved from True North: https://truenorthimpact.com/publications/aging-populations-venture-capital
6. Shestakova, N. N., Djanelidze, M. G., & Skvortsova, M. B. (2022). *AgeTech innovations as an instrument for older population inclusion.* Paper presented at the National Conference on Sustainable Development (ESG).
7. Dalmer, N., Ellison, K., Katz, S., & Marshall, B. (2022). Ageing, embodiment and datafication: Dynamics of power in digital health and care technologies. *International Journal of Ageing and Later Life, 15*(2), 77-101.
8. Scott, A. J. (2021). The longevity economy. *The Lancet Healthy Longevity, 2*(12), e828-e835. https://doi.org/10.1016/S2666-7568(21)00250-6
9. Anderson, M., & Perrin, A. (2017). Tech adoption climbs among older adults. Retrieved from https://www.pewresearch.org/internet/2017/05/17/tech-adoption-climbs-among-older-adults/

5 Public and Private Funding in Aging Research and Services: A Collaborative Imperative

5.1 Introduction

The global aging population is reshaping societies at an extraordinary pace. According to the World Health Organization (WHO), the number of people aged 60 and over is expected to more than double by 2050, reaching over 2 billion [1]. This demographic shift has profound implications for healthcare, social security, and economic systems. Aging-related conditions such as dementia, cardiovascular diseases, and osteoarthritis are placing immense pressure on healthcare infrastructure [2].

In Canada, approximately 18% of the population is aged 65 or older, a figure expected to rise to nearly 25% by 2030. Globally, countries like Japan, Germany, and the United States face similar demographic shifts, each presenting unique opportunities and challenges. Addressing these requires an integrated approach that combines public funding for accessibility and infrastructure with private funding for innovation and scalability [3].

5.2 The Role of Public Funding

Public funding forms the backbone of advancements in aging research and services, addressing systemic challenges while promoting equity. Globally, public funding agencies have been pivotal in supporting large-scale aging research.

In Canada, the Canadian Institutes of Health Research (CIHR) provides substantial funding through its Institute of Aging (CIHR-IA), focusing on dementia, caregiving, and geroscience. In 2020 alone, CIHR allocated over $60 million to aging-related projects. Similarly, the Natural Sciences and Engineering Research Council (NSERC) supports technological innovations, such as assistive robotics and health-monitoring devices.

A. Mihailidis, *The Age of AgeTech*, Synthesis Lectures on Technology and Health,
https://doi.org/10.1007/978-3-032-04886-8_5

In the United States, the National Institute on Aging (NIA), a division of the National Institutes of Health (NIH), had a budget of $3.54 billion in 2020, supporting a wide range of research on Alzheimer's disease and longevity interventions. The European Union, through its flagship research program Horizon 2020, allocated €850 million to aging-related initiatives, fostering interdisciplinary research and technological development. Japan's Ministry of Health, Labour and Welfare also invests in aging research, particularly focusing on robotics and assistive technologies to address eldercare challenges.

Public funding is not limited to research but also extends to the development of essential infrastructure for aging populations. For example, Canada's New Horizons for Seniors Program allocated $60 million in 2021 to community-based projects aimed at enhancing social inclusion and well-being [4]. In Japan, public investments have enabled the development of smart home initiatives and robotic caregiver technologies to address workforce shortages in eldercare.

Equitable access to aging services is another critical goal of public funding. Canada's Healthy Aging initiative through the Public Health Agency of Canada promotes the development of age-friendly communities and addresses disparities in healthcare access. Similarly, the WHO's Healthy Aging strategy emphasizes equitable resource allocation, ensuring that advancements in aging research and services benefit all populations [1].

5.3 Private Sector Contributions

The private sector complements public funding by driving innovation, accelerating commercialization, and addressing unmet needs in aging-related services. Corporate investments in aging innovation have been transformative.

In Canada, companies like PointClickCare and AlayaCare are at the forefront of digital health. PointClickCare's cloud-based platform enhances long-term care facility management, while AlayaCare develops software solutions for home healthcare agencies. In the United States, Google's Calico Labs focuses on the biology of aging, exploring pathways to extend health spans. IBM Watson Health leverages AI to deliver personalized healthcare solutions for seniors [5].

Venture capital investment plays a critical role in advancing AgeTech. Canadian startups like Careteam Technologies, which offers care coordination software, and Kinova Robotics, known for assistive robotic arms, have secured significant funding to expand their services. Internationally, companies like Honor Technology, a U.S.-based home care platform, have raised over $325 million to modernize caregiving services. ElliQ, an Israeli-developed robotic companion for seniors, addresses issues such as loneliness and mental health challenges, illustrating the diversity of AgeTech innovations [6].

Philanthropy also plays a crucial role in bridging funding gaps, particularly for high-risk, high-reward research. In Canada, the Baycrest Foundation supports cognitive and brain health research through initiatives like the Baycrest Academy for Research and

Education. In the United States, the Gates Foundation backs aging-related global health initiatives, particularly in low-income countries, showcasing the potential for philanthropy to address disparities in aging research and services [7].

5.4 Collaborative Models of Funding

Public–private partnerships (PPPs) are a powerful model for maximizing the strengths of both sectors, fostering innovation, and scaling solutions.

Successful collaborations highlight the potential of these models. In Canada, AGE-WELL, funded through the Networks of Centres of Excellence (NCE) program), collaborates with over 600 partners, including startups, universities, and non-profits. Its projects span smart home technologies, AI caregiving tools, and health data analytics [8].

Internationally, Japan's Society 5.0 initiative combines public and private efforts to create a super-smart society, with solutions such as IoT-enabled homes and AI-powered diagnostics. The UK's Accelerated Access Collaborative streamlines regulatory pathways for health technologies, accelerating their deployment in aging-related healthcare.

Despite their advantages, collaborative funding models are not without challenges. Misaligned incentives often create friction, as public entities prioritize equity while private firms focus on profitability. Aligning these goals requires clear contractual agreements and shared objectives. Regulatory barriers also pose significant hurdles. For instance, Canada's regulatory framework for medical devices often involves lengthy approval processes through Health Canada, delaying innovation. Ethical considerations, such as data privacy and equitable resource allocation, are additional concerns that require transparency and accountability to maintain public trust.

5.5 Strategies for Enhancing Funding Efficiency

Improving the effectiveness of public and private funding requires targeted strategies. Expanding AGE-WELL's mandate and regional reach, for instance, could enhance its capacity to support diverse AgeTech projects across Canada. Establishing provincial hubs would address regional disparities and foster innovation at a local level. Encouraging impact investing is another promising strategy. Impact investors, who prioritize financial returns alongside social outcomes, are ideal partners for aging-related initiatives. Canadian pension funds, such as the Canada Pension Plan Investment Board (CPPIB), have already begun incorporating impact investing principles, signaling a shift toward more socially responsible investment practices.

Strengthening academic-industry partnerships can also drive progress. Programs like Mitacs, which provide funding and training for Canadian graduate students working on industry-led aging projects, foster innovation through collaboration between academia and the private sector. Creating regulatory sandboxes is another practical approach. These controlled environments allow companies to trial new technologies without full regulatory compliance, expediting approval processes. The OECD has recommended regulatory sandboxes as an effective tool for fostering innovation.

5.6 Future Directions

As the global aging population continues to grow, the demand for innovative solutions will intensify. Emerging trends in personalized medicine, digital health, and AI offer promising avenues for investment. Canada's leadership in collaborative funding models and international partnerships positions it to play a pivotal role in addressing aging-related challenges. By fostering a culture of innovation and collaboration, stakeholders can ensure that aging populations are supported with dignity and care.

5.7 Conclusion

Public and private funding streams are indispensable for addressing the opportunities and challenges posed by aging populations. Canada's initiatives, such as CIHR's aging research programs, AGE-WELL's collaborative network, and corporate investments in AgeTech, exemplify the potential of coordinated efforts. On the international stage, examples like Japan's Society 5.0 and the UK's Accelerated Access Collaborative underscore the value of cross-sector partnerships.

By aligning incentives, streamlining regulations, and fostering dialogue, stakeholders can ensure sustainable and effective funding solutions for aging research and services, setting a global standard for innovation and equity.

References

1. WHO. (2015). *World report on aging and health.* Retrieved from Geneva: https://www.who.int/publications/i/item/9789241565042/
2. Queiroz, L. L. C. d. S., & Boente, D. R. (2023). Financing Public Health Expenditure in an Aging Society: The Role of Public-Private Partnerships in Health. *New Challenges in Accounting and Finance, 9*(1), 1–12. https://doi.org/10.32038/NCAF.2023.09.01
3. Remillard, E. T., Campbell, M. L., Koon, L. M., & Rogers, W. A. (2022). Transportation challenges for persons aging with mobility disability: Qualitative insights and policy implications. *Disability and Health Journal, 15*(1, Supplement), 101209. https://doi.org/10.1016/j.dhjo.2021.101209
4. Government of Canada. (2024). About the New Horizons for Seniors Program. Retrieved from https://www.canada.ca/en/employment-social-development/programs/new-horizons-seniors.html

5. Li, X., Zhang, S., & Meng, F. (2022). Operation mode selection of public-funded private-run elderly care agencies based on triangular intuitionistic fuzzy preference relations group decision-making. *Computational and Applied Mathematics, 41*(8), 346. https://doi.org/10.1007/s40314-022-02050-7
6. Ng, A. W., Leung, T. C. H., & Ka Tat Tsang, A. (2020). Social Enterprise for Elderly Housing: Policy for Accountability and Public-Private Responsible Financing. *Journal of Population Ageing, 13*(3), 365–384. https://doi.org/10.1007/s12062-018-9235-5
7. Yeh, C.-Y., Cheng, H., & Shi, S.-J. (2018). Public–private pension mixes in East Asia: Institutional diversity and policy implications for old-age security. *Ageing and Society, 40*(3), 604–625. https://doi.org/10.1017/S0144686X18001137
8. AGE-WELL. (2021). Networks of Centres of Excellence: Aging and Technology Initiatives in Canada. https://www.agewell-nce.ca

6 Sustainable Funding for Aging Services: Securing the Future of Elder Care

6.1 Introduction

The aging population represents one of the most profound demographic shifts of the twenty-first century. By 2050, the proportion of people aged 65 and older is expected to reach nearly 25% in developed nations, with similar trends in developing regions [8]. This demographic transformation presents significant financial opportunities and challenges, as older adults require more healthcare services, long-term care, and social support [7].

Current funding mechanisms need to be revised to explore these growing needs sustainably. Public systems face budgetary constraints, while private funding often excludes low-income and vulnerable populations. This essay delves into the opportunities and challenges of funding aging services, evaluates current models, and proposes innovative strategies for ensuring long-term sustainability.

6.2 The Economic Impact of Aging Populations

Healthcare utilization increases significantly with age. Older adults are more likely to have chronic conditions such as cardiovascular disease, diabetes, and dementia, all of which require long-term management.

In Canada, individuals aged 65 and older account for nearly 45% of public healthcare expenditures, despite representing only 18% of the population [3]. Similar trends are observed globally, with countries like Japan and Germany allocating more than 50% of healthcare budgets to elder care.

Long-term care (LTC) services, including nursing homes and home-based care, constitute significant aging-related costs. In the United States, LTC spending reached $235 billion in 2020, with Medicaid covering over half of these expenses [4]. In Canada, public

A. Mihailidis, *The Age of AgeTech*, Synthesis Lectures on Technology and Health,
https://doi.org/10.1007/978-3-032-04886-8_6

LTC spending exceeded $22 billion in 2021, with provinces and territories shouldering most costs [3].

Unpaid caregiving by family members contributes significantly to the economy and places financial and emotional burdens on caregivers. In Canada, the estimated value of unpaid caregiving was $26 billion in 2020. Addressing these costs requires a holistic funding approach integrating formal and informal care systems [5].

6.3 Challenges in Current Funding Models

Public funding, while essential, is increasingly strained by rising costs and competing priorities. Aging services often face budget cuts during economic downturns, leaving vulnerable populations underserved.

In Canada, provincial healthcare systems operate under tight fiscal constraints, leading to regional disparities in service quality, especially in northern and Indigenous communities [3].

Private funding models, such as private insurance and out-of-pocket payments, are inaccessible to many older adults. High premiums and deductibles deter enrollment, while low-income individuals are often excluded entirely.

Aging services are funded through a patchwork of sources, including public programs, private insurance, and philanthropy. This fragmentation complicates coordination and leads to inefficiencies, such as duplication of services and gaps in coverage.

Finally, socioeconomic and geographic disparities exacerbate inequities in access to aging services. Rural areas often lack adequate infrastructure, while low-income populations face barriers to affordable care.

6.4 Current Models of Funding Aging Services

6.4.1 Publicly Funded Models

There are several examples of publicly funded models around the world that have been implemented to support older people.

- **Canada**: Medicare covers essential healthcare services, while provincial programs fund long-term care and home care services. However, these programs are underfunded, leading to long wait times and limited availability [3].
- **Japan**: The Long-Term Care Insurance (LTCI) system covers 90% of eligible care costs and is financed through public funds and mandatory premiums for individuals aged 40 and older.

- **Nordic countries (Denmark and Sweden)**: Offer universal access to LTC services funded through general taxation, ensuring equity and affordability.

6.4.2 Private Funding Models

In addition to publicly funded models, private models also support older people:

- **United States**: Medicaid and Medicare cover most aging-related costs, but private LTC insurance plays a supplemental role. Enrollment in private LTC insurance remains low due to high premiums and limited coverage options [4].
- **Germany**: Uses hybrid systems where public funds cover essential services, and private insurance provides additional benefits.
- **Non-profits and philanthropy**: play a crucial role in filling funding gaps. In Canada, initiatives like the Baycrest Foundation support cognitive health research and aging-related education [2].

6.5 Strategies for Sustainable Funding

Public funding remains the backbone of aging services, and increasing allocations to aging programs is essential for ensuring universal access and equity.

- **Public funding mechanisms**: Implementing higher taxes on high-income earners and establishing dedicated aging funds, similar to Japan's LTCI system, to provide predictable, sustainable financing.
- **Private investment**: Encouraging impact investing can direct capital to aging services with measurable social and financial returns, as seen in initiatives supported by the Rockefeller Foundation [6].
- **Public–private partnerships**: These enhance resource efficiency and innovation, exemplified by AGE-WELL's collaborations with tech companies in Canada [1].
- **Value-based care**: Implementing outcome-based contracts and models like Accountable Care Organizations (ACOs) can reduce unnecessary expenditures and improve patient outcomes [4].
- **Leveraging technology**: Telemedicine improves accessibility and lowers costs, particularly in rural areas, while AI-based data analytics can identify at-risk populations and optimize resource allocation.
- **Supporting informal caregivers**: Offering tax incentives and expanding respite care services can alleviate caregiver burdens, prevent burnout, and improve outcomes for older adults.

6.6 Case Studies

Japan's Long-Term Care Insurance (LTCI) system, introduced in 2000, highlights the effectiveness of dedicated funding, covering most eldercare costs through mandatory premiums and public subsidies, thereby improving access to care and reducing the financial burden on families. Similarly, Denmark's universal, tax-funded long-term care (LTC) system ensures that all citizens have access to high-quality services, placing a particular emphasis on home care to reduce reliance on institutional facilities and promote aging in place. In Canada, AGE-WELL integrates public and private funding to advance AgeTech innovation, supporting projects that develop smart home technologies, telehealth platforms, and robotic caregiving tools, all of which improve efficiency and sustainability in aging services. In the United States, value-based care models introduced by the Centers for Medicare & Medicaid Services (CMS) demonstrate a shift toward preventative care and chronic disease management, with programs like Medicare Advantage incentivizing providers to enhance efficiency and focus on long-term outcomes.

6.7 Future Directions

Sustainable funding for aging services requires a multifaceted approach. Emerging trends, such as personalized medicine and AI-driven care, offer promising avenues for innovation. International collaboration and knowledge sharing will be critical in exploring global aging opportunities and challenges.

6.8 Conclusion

Sustainable funding is essential to meet the growing demands of aging populations. Public funding must remain the foundation of aging services, supported by private investment, value-based care models, and innovative technologies.

By learning from international examples and leveraging collaborative frameworks, stakeholders can create a resilient and equitable system for elder care. Addressing the financial opportunities and challenges of aging is not just a necessity, but an opportunity to improve the quality of life for older adults worldwide.

References

1. AGE-WELL. (2021). Networks of Centres of Excellence: Aging and Technology Initiatives in Canada
2. Baycrest. (2021). Cognitive Health Research and Education Initiatives in Canada

3. Canadian Institute for Health Information. (2024). National health expenditure trends. Retrieved from https://www.cihi.ca/en/national-health-expenditure-trends
4. CMS. (2021). Medicare and Medicaid's Role in Long-Term Care Funding
5. Fang, E. F., Xie, C., Schenkel, J. A., Wu, C., Long, Q., Cui, H.,... Woo, J. (2020). A research agenda for ageing in China in the 21st century (2nd edition): Focusing on basic and translational research, long-term care, policy and social networks. *Ageing Research Reviews, 64*, 101174. https://doi.org/10.1016/j.arr.2020.101174
6. Rockefeller Foundation. (2020). Impact Investing in Aging Services
7. Rudnicka, E., Napierała, P., Podfigurna, A., Męczekalski, B., Smolarczyk, R., & Grymowicz, M. (2020). The World Health Organization (WHO) approach to healthy ageing. *Maturitas, 139*, 6–11. https://doi.org/10.1016/j.maturitas.2020.05.018
8. United Nations. (2024). World Population Prospects 2024. Retrieved from https://www.un.org/development/desa/pd/world-population-prospects-2024

7 Why AgeTech Matters: Transforming the Future of Aging

7.1 Introduction

Aging is one of the most significant demographic phenomena of our time. The world is on the brink of a dramatic shift in age structure, with older adults becoming an increasingly dominant population segment. In Canada, for example, 18% of the population is 65 and older, which is projected to rise to 25% by 2030. Globally, aging populations bring opportunities and challenges to healthcare systems, economies, and social structures [7].

AgeTech—technology tailored to meet the needs of older adults—is emerging as a vital tool to navigate these opportunities and challenges. AgeTech encompasses many innovations, from assistive devices and smart home technologies to telemedicine platforms and artificial intelligence (AI) tools for eldercare. By exploring the unique needs of aging populations, AgeTech promotes autonomy, enhances healthcare delivery, and improves the quality of life for seniors. This essay examines why AgeTech matters, emphasizing its transformative potential and providing international examples of its impact.

7.2 The Demographic Context of Aging

The global population aged 65 and older is projected to double from 703 million in 2019 to 1.5 billion by 2050, a trend driven by increased life expectancy and declining fertility rates [8]. Although this demographic shift is most pronounced in developed countries, developing nations are also experiencing rapid aging [6].

Canada reflects this global trend, as seniors represent the fastest growing demographic while Japan, where 28% of the population is over 65, offers a stark example of the

A. Mihailidis, *The Age of AgeTech*, Synthesis Lectures on Technology and Health,
https://doi.org/10.1007/978-3-032-04886-8_7

societal implications of an aging population, including workforce shortages and escalating healthcare costs.

The economic and social ramifications of population aging affect multiple sectors:

- Healthcare systems face greater demands due to the rising prevalence of chronic conditions such as dementia, arthritis, and cardiovascular diseases.
- Social support systems, including pensions and social security programs, come under strain.
- Workforce dynamics shift, as fewer working-age individuals lead to reduced economic productivity and higher dependency ratios.

In response to these challenges, AgeTech presents innovative solutions that not only meet the complex needs of aging populations but also help alleviate broader societal pressures [4].

7.3 Healthcare Innovations in AgeTech

Telemedicine and remote monitoring technologies have revolutionized healthcare access for older adults, particularly those in rural or underserved areas.

- Wearable sensors and health trackers enable continuous health management, reducing both hospital visits and overall costs.
- In Canada, companies like AlayaCare offer virtual care platforms tailored to seniors, allowing caregivers to monitor vital signs, medication adherence, and mobility in real time for more effective, personalized care.
- In Japan, telemedicine platforms like Dokodemo Doctor help address physician shortages in rural areas, connecting seniors with healthcare professionals through mobile devices [3].

Beyond telehealth, AI-driven diagnostics and treatment tools such as IBM Watson Health analyze patient data to detect early signs of cognitive decline or chronic diseases.

- In Canada, AGE-WELL focuses on AI caregiving systems to enhance eldercare efficiency and ease the burden on healthcare workers.
- Robotic assistance also plays a critical role: In Japan, investments in robotic caregivers like PARO, a therapeutic robot for dementia patients, and Robear, a robot designed to assist with patient lifting and transferring, illustrate how technology can support both physical and cognitive well-being.
- In the United States, service robots like Moxi help healthcare staff by handling routine tasks, freeing them to provide more direct, individualized patient care.

7.4 Enhancing Social Inclusion and Mental Health

Combating loneliness and isolation is a key focus of AgeTech, as social isolation significantly affects older adults' mental health, contributing to issues like depression and anxiety [7].

- In Canada, the Telus Health Companion platform provides seniors with virtual social networks and telehealth services, helping mitigate loneliness and ensure continuous support.
- In Europe, ElliQ, a social robot developed in Israel, interacts conversationally with older adults, offering reminders and cognitive games to improve their mental well-being [1].

Beyond fostering connection, promoting digital literacy is essential for bridging the technology gap among seniors [2].

- In Singapore, the Seniors Go Digital program organizes workshops that teach older adults how to navigate smartphones and digital platforms, empowering them to stay connected and access vital services.

7.5 Economic Benefits of AgeTech

AgeTech not only improves care quality but also has tangible economic benefits by reducing healthcare costs and fostering job growth [5].

- In the United States, a Centers for Medicare & Medicaid Services (CMS) study demonstrated that remote patient monitoring reduced hospital readmissions by 25% for seniors with chronic diseases, resulting in substantial cost savings and improved patient outcomes.
- In Canada, AGE-WELL's innovation hubs and academic-industry partnerships nurture startups that develop new technologies, generating skilled jobs and attracting investment.

7.6 Challenges and Barriers to Adoption

While AgeTech offers transformative possibilities, several challenges must be addressed to ensure its equitable, ethical, and inclusive adoption. One significant hurdle is the digital divide: many older adults struggle with limited digital literacy, lack reliable internet access, or simply cannot afford the necessary devices. Overcoming this gap requires targeted educational initiatives, such as workshops at community centers or libraries, dedicated training sessions led by non-profits, and simplified user interfaces, as well as affordable pricing models or device subsidies. Ethical considerations further complicate the landscape, as AgeTech solutions often involve sensitive personal data and automated caregiving tasks. Transparent policies, stringent data privacy regulations, and human-centered design principles can help maintain user trust, ensuring that emerging technologies support well-being without undermining autonomy or compassion. Additionally, cost and accessibility remain pressing concerns, since expensive devices and subscription fees can exclude low-income seniors from benefiting fully. Public funding, means-tested subsidies, strategic partnerships with philanthropic organizations, and the promotion of open-source solutions are all vital steps toward ensuring that no older adult is left behind in this digital revolution. Ultimately, by actively addressing these issues, stakeholders can create an AgeTech environment that empowers all seniors, protects their rights and dignity, and ensures broader societal gains.

7.7 Policy and Advocacy for AgeTech

Governments and organizations around the world have a crucial role to play in shaping a supportive ecosystem for AgeTech, implementing policies and funding initiatives that catalyze innovation and ensure equitable access. In Canada, for instance, the Canadian Institutes of Health Research (CIHR), through its Institute of Aging, supports robust research programs that fuel advancements in eldercare technologies and evidence-based interventions. Simultaneously, community-based initiatives like the New Horizons for Seniors Program provide funding for local projects that empower older adults, promote social inclusion, and foster digital literacy, ultimately laying the groundwork for sustainable AgeTech adoption. On a global scale, the World Health Organization's Healthy Aging initiative urges member states to incorporate AgeTech into their national healthcare strategies, emphasizing that thoughtfully integrated technologies can improve care coordination, enhance disease prevention, and support the autonomy of older individuals. By prioritizing AgeTech at national and international levels, decision-makers can ensure that technological progress aligns with public health goals, social well-being, and economic sustainability, ultimately creating an environment where older adults can thrive.

7.8 Case Studies: Success Stories in AgeTech

Japan's approach to AgeTech exemplifies how cutting-edge robotics can help address pressing challenges in elder care. Robotic caregivers like Robear, capable of lifting and transferring patients, and PARO, a therapeutic seal-like robot designed to comfort individuals with dementia, serve as innovative solutions to workforce shortages while enhancing care quality and patient well-being. In Canada, the AGE-WELL network illustrates the power of collaboration, bringing together over 200 organizations, ranging from startups and established tech firms to research institutions and community groups, to drive AgeTech advancements. These efforts have yielded impactful projects like smart home monitoring systems, virtual care platforms, and AI-driven diagnostic tools that streamline care delivery and improve outcomes. In the United States, the expansion of telehealth coverage by the Centers for Medicare & Medicaid Services during the COVID-19 pandemic showcases AgeTech's adaptability and responsiveness to emerging needs. By broadening access to remote consultations and digital health tools, policymakers not only mitigated immediate healthcare constraints but also laid the groundwork for more efficient, patient-centric care models. Together, these international examples underscore the vast potential of AgeTech to enhance eldercare experiences, reduce system strain, and respond nimbly to shifting demographic and healthcare landscapes.

7.9 Conclusion

AgeTech is not merely a response to the opportunities and challenges of aging populations—it is a transformative force reshaping how societies approach aging.

From telemedicine in Canada to robotic caregivers in Japan, AgeTech is paving the way for a future where aging is met with dignity, innovation, and resilience. To fully realize its potential, policymakers, businesses, and communities must collaborate to invest in and scale AgeTech solutions, ensuring they are accessible to all.

References

1. Bryanton, O., Dragomir, G., Mann, J., & Moulton, M. (2023). Older Adults' Perspectives on Ethical Issues Related to AgeTech. *Innovation in Aging, 7*(Supplement_1), 312–312
2. Coughlin, J. F. (2020). The Fourth Wave of Technology and Aging: Policy Innovation to Ensure Equity and Inclusion. *Public Policy & Aging Report, 30*(4), 138–141. https://doi.org/10.1093/ppar/praa032
3. Gomez-Hernandez, M. (2024). Industry visions of technology for older adults: A futures anthropology perspective. *Journal of Aging Studies, 70*, 101248. https://doi.org/10.1016/j.jaging.2024.101248
4. Lepore, M. J., & McMullen, T. (2024). Public Policy and Aging Research: After 50 Years of the National Institute on Aging. *Public Policy & Aging Report, 34*(4), 111–113. https://doi.org/10.1093/ppar/prae025
5. Schafer, D., & Rutley, J. (2021). *Aging Populations: The Role Of Venture Capital In Supporting The Industry.* Retrieved from True North: https://truenorthimpact.com/publications/aging-populations-venture-capital
6. Shestakova, N. N., Djanelidze, M. G., & Skvortsova, M. B. (2022). *AgeTech innovations as an instrument for older population inclusion.* Paper presented at the National Conference on Sustainable Development (ESG)

7. Sixsmith, A., & Cosco, T. D. (2024). Technology, aging, and mental health: opportunities and challenges. *Aging & Mental Health, 28*(4), i–iii. https://doi.org/10.1080/13607863.2024.2320132
8. United Nations. (2024). World Population Prospects 2024. Retrieved from https://www.un.org/development/desa/pd/world-population-prospects-2024

The Rise of AgeTech: Key Innovations, Assistive Devices, and Smart Home Solutions

8

8.1 Introduction

The global demographic landscape is undergoing a profound transformation. As medical advances, improved sanitation, better nutrition, and enhanced public health measures collectively extend human life expectancy, societies worldwide are witnessing a rapid increase in the proportion of older adults [4].

According to projections by the [4], the number of people aged 60 and above is expected to surpass two billion by 2050. While this aging trend is most frequently discussed in the context of high-income nations, such as Japan, many European countries, and parts of North America, this demographic shift is by no means confined to the developed world. Low- and middle-income countries, benefiting from declining fertility rates and improving healthcare infrastructures, are also seeing significant increases in their older populations [1].

This global "gray wave" has critical implications not only for healthcare systems but also for economic productivity, labor markets, social policies, community design, and cultural norms related to aging.

In many countries, the traditional models of eldercare, often reliant on family caregivers, community support, and limited institutional care, are strained by the sheer scale of the demographic shift. As older adults live longer, they face a higher likelihood of developing chronic conditions, such as diabetes, heart disease, and various forms of dementia. These conditions demand continuous, often complex care that extends well beyond occasional physician visits.

Furthermore, the potential for social isolation, loneliness, and reduced mobility increases as seniors age, especially if their local environments are not designed with their needs in mind [2]. Without innovation and adaptation, healthcare systems and social structures may struggle to provide sustainable, high-quality care for aging populations.

A. Mihailidis, *The Age of AgeTech*, Synthesis Lectures on Technology and Health,
https://doi.org/10.1007/978-3-032-04886-8_8

In response to these emerging needs and challenges, AgeTech has arisen as a promising field of innovation. "AgeTech" can be broadly defined as the application of technology, encompassing digital tools, smart devices, robotics, artificial intelligence (AI), telemedicine, and home automation, to improve the quality of life for older adults.

Beyond enhancing health outcomes, AgeTech solutions aim to promote independence, reduce the burden on family caregivers and healthcare workers, alleviate social isolation, and ensure that societies can adapt to the realities of an aging demographic. Far from treating older individuals as passive recipients of care, AgeTech positions them as active participants in their health management and daily life routines.

This shift in perspective is evident through a wide array of innovations: wearable health devices that empower seniors to track their own vital signs; telehealth platforms that connect them remotely to physicians and nurses; cognitive health technologies that preserve mental acuity; assistive mobility aids that restore independence; advanced hearing and vision technologies; and smart home solutions that allow for aging in place with greater safety and comfort. Moreover, these technologies do not exist in isolation. They often integrate into broader ecosystems, smart homes, telehealth networks, community support services, and policy frameworks, that support seniors throughout the aging journey.

International examples underscore the versatility and impact of AgeTech. Countries like Japan, facing one of the oldest populations in the world, are pioneering the use of robotic caregivers and AI-driven solutions to offset workforce shortages. Canada's AGE-WELL network has become a global leader in fostering AgeTech innovation, catalyzing partnerships across industry, academia, and community organizations. The United States, propelled by private sector leadership and responsive policy shifts, has expanded telemedicine coverage and integrated AgeTech into both clinical and home settings. European countries have long prioritized universal design principles, ensuring that assistive devices and inclusive environments support seniors' autonomy and dignity.

However, AgeTech's rise also brings challenges. Issues related to the digital divide, some older adults may lack the skills, access, or comfort to use new technologies, must be addressed through education and affordable access. Ethical concerns around data privacy, automation's impact on human interaction, and equitable distribution of AgeTech resources must be carefully considered. Policymakers, industry leaders, researchers, and advocacy groups need to collaborate, ensuring that AgeTech solutions remain inclusive, user-centered, and ethically grounded.

This comprehensive overview will delve into the key innovations in technology for seniors, examine the diverse range of assistive devices, explore smart home solutions, consider international perspectives, and reflect on the path forward. By mapping out the landscape of AgeTech and examining its multiple facets, we can better understand how societies worldwide are leveraging these innovations to achieve healthier, more sustainable, and more dignified aging experiences.

8.2 Key Innovations in Technology for Seniors

Within the rapidly evolving AgeTech ecosystem, some of the most visible and impactful innovations focus on directly enhancing seniors' day-to-day health management, communication with healthcare providers, and mental well-being. Below, we examine three critical domains of technological advancement: wearable health devices, telehealth platforms, and cognitive health technologies.

8.2.1 Wearable Health Devices

Wearable technologies have become ubiquitous in recent years, with devices like smartwatches and fitness trackers appealing to a broad consumer base. For older adults, however, wearables represent far more than a fitness fad. They are tools that can literally help save lives, manage chronic conditions, and foster independence. Modern wearable health devices can track a variety of health indicators: heart rate, blood pressure, sleep quality, activity levels, blood glucose levels, and even subtle changes in gait or balance that may signal emerging mobility issues.

Common consumer devices like Fitbit and the Apple Watch have demonstrated that older adults can be empowered to understand their own health metrics. For instance, a senior who notices a persistent elevation in heart rate or decreased sleep quality may be prompted to consult a physician earlier, potentially detecting problems at a more treatable stage. Apple has incorporated features like fall detection and atrial fibrillation monitoring, which can be lifesaving if an older adult experiences an unexpected medical event.

Beyond consumer brands, specialized healthcare-focused companies cater specifically to older populations with more advanced medical-grade devices. Canada's Biotricity, for example, leads in remote cardiac monitoring. By enabling real-time heart health oversight, Biotricity's products help reduce emergency department visits and hospital readmissions. Physicians can access patients' data remotely and adjust treatment plans accordingly, making cardiac care more proactive, personalized, and cost-effective.

These technologies collectively reduce the strain on healthcare systems by shifting from reactive to proactive care. Instead of only treating acute incidents after they occur, wearable devices allow for the early detection of irregularities. This early intervention can translate to fewer hospitalizations, improved patient outcomes, and a greater sense of agency for older adults who can, to some degree, manage their own health destinies.

8.2.2 Telehealth Platforms

Telehealth has expanded rapidly, especially after the onset of the COVID-19 pandemic. For seniors, telehealth is particularly valuable: it overcomes common barriers such as mobility limitations, transportation challenges, and the difficulty of scheduling and traveling to in-person appointments. Through secure video conferencing, online chat, and remote diagnostic tools, older adults can consult with healthcare professionals without leaving their homes.

In the United States, platforms like Teladoc Health offer a range of services, connecting patients to physicians, specialists, mental health professionals, and more. This reduces not only travel-related hardships but also the time spent in waiting rooms, exposure to potential infections, and the stress of navigating unfamiliar clinical environments. Physicians can review patient data from wearables and electronic health records, providing personalized recommendations and follow-ups.

Similarly, Canada's Maple has been instrumental in broadening access to telehealth solutions. Such platforms serve rural communities, individuals with mobility constraints, and those who simply prefer the convenience of remote care. Telehealth is also critical for managing chronic conditions over the long term. A senior with diabetes, for example, can regularly report blood glucose measurements to a care team, receive timely adjustments to their regimen, and avoid complications that could lead to emergency care.

8.2.3 Cognitive Health Technologies

As longevity increases, so does the risk of dementia, mild cognitive impairment, and other cognitive challenges. Cognitive health technologies aim to preserve and even enhance cognitive function, recognizing that maintaining mental acuity is essential for overall quality of life, independence, and social engagement.

Israel's ElliQ, an AI-powered robotic companion, represents a novel approach to supporting cognitive health. ElliQ is not a mere gadget; it is designed to engage older adults in conversation, offer reminders for medication or appointments, prompt participation in cognitive games, and encourage activities that keep the mind active. Interacting with ElliQ can help reduce feelings of loneliness, known to exacerbate cognitive decline, and create a sense of routine and structure that supports brain health.

In Canada, AGE-WELL's research on cognitive care solutions demonstrates how AI tools can monitor subtle changes in language use, task completion times, or memory tests. These insights can flag emerging cognitive issues, allowing healthcare providers and caregivers to intervene early. Such tools may also support personalized care plans, recommending activities tailored to each individual's cognitive profile, and encouraging social interactions through digital communities or interest-based activities.

8.3 Assistive Devices for Seniors

While wearable devices, telehealth, and cognitive tools address health monitoring and mental wellness, other facets of AgeTech concentrate on enabling physical independence and compensating for sensory impairments. Mobility aids and assistive devices for hearing and vision are particularly crucial, as they allow older adults to remain engaged in their communities, maintain personal autonomy, and continue enjoying life's fundamental activities.

8.3.1 Mobility Aids

Mobility often declines with age due to conditions like arthritis, muscle weakness, or balance disorders. Reduced mobility can lead to isolation, as older adults become hesitant to leave their homes, participate in community events, or perform daily errands. AgeTech offers innovative solutions to these challenges.

In Japan, Cyberdyne's exoskeletons exemplify how wearable robotics can restore mobility and strength. These devices detect the user's muscle signals and provide mechanical support, enabling individuals with limited mobility to stand, walk, and navigate their environments more easily. While still emerging technologies, exoskeletons hold substantial promise for both personal and institutional settings, reducing the physical burden on caregivers and encouraging seniors to remain active longer.

In the Netherlands, Rollz Motion has integrated a walker and a wheelchair into a single adaptable device. This dual-function solution can be pushed as a walker or converted into a wheelchair as needed, promoting versatility during outings. Such flexible, thoughtfully designed aids empower older adults to remain active participants in their communities rather than passive recipients of mobility assistance.

8.3.2 Hearing and Vision Devices

Hearing and vision loss are common in older adulthood and can significantly impact communication, safety, and overall well-being. AgeTech focuses on restoring sensory capabilities or providing compensatory solutions that ensure older adults can continue living independently and meaningfully engaging with the world.

Denmark's Oticon leverages AI to improve speech clarity in hearing aids. Traditional hearing aids often amplify all sounds equally, making it challenging to discern speech in noisy environments. By using sophisticated algorithms to filter and enhance speech signals, modern hearing aids help seniors follow conversations, watch television, and participate in community gatherings with greater ease.

Canada's eSight provides an innovative solution for individuals who are legally blind. Their electronic glasses combine high-definition camera technology, image processing, and intuitive controls to restore functional vision. Users can read, recognize faces, watch their grandchildren play, and appreciate their surroundings in ways that were previously impossible. Such assistive devices not only improve quality of life but also promote independence and self-confidence.

8.4 Smart Home Solutions

As populations age, the concept of "aging in place", remaining in one's own home and community as long as possible, has gained traction. Aging in place is preferable for many seniors who wish to maintain familiarity, independence, and control over their daily routines. Smart home technologies, powered by the Internet of Things (IoT), artificial intelligence, and home automation systems, offer practical tools that enhance safety, convenience, and care coordination.

8.4.1 IoT-Enabled Smart Homes

IoT-enabled smart homes integrate various sensors, devices, and appliances into a cohesive system that older adults can manage through intuitive interfaces or voice commands. Voice assistants like Amazon's Alexa and Google Nest devices allow seniors to adjust thermostats, turn lights on and off, or play their favorite music simply by speaking. They can also set medication reminders, check the weather, or order groceries online, reducing the need for complicated navigation on smartphones or computers.

In Japan, Panasonic's smart home solutions go beyond convenience to prioritize safety. Motion sensors, fall detectors, and automated lighting systems reduce the risk of accidents at home. For instance, if a sensor detects an unusual lack of movement during the day, it can alert a caregiver or family member to check in. Similarly, automated lighting may switch on when someone gets out of bed at night, preventing a stumble in the dark.

8.4.2 Fall Detection Systems

Falls represent a leading cause of injury, hospitalization, and loss of independence among seniors. More than a physical hazard, falls can also erode self-confidence, leading individuals to become more sedentary and isolated. AgeTech addresses this challenge with sophisticated fall detection systems.

Canada's VitaliWear provides wearable devices equipped with AI that can detect sudden changes in orientation, acceleration, or impact indicative of a fall. Once a fall is detected, the device immediately alerts caregivers or designated emergency contacts. Rapid response can mean the difference between a minor injury and a life-threatening situation. Similarly, in the United States, Philips Lifeline offers emergency response systems specifically tailored to older adults. These systems can be linked to professional monitoring centers, ensuring assistance is dispatched promptly.

8.4.3 Home Healthcare Robotics

Robotics has enormous potential to assist with daily tasks and alleviate caregiver burdens. While robots capable of comprehensive eldercare are still in development, existing examples illustrate how robotic assistants can improve both practical and emotional aspects of care.

In Japan, Robear, a gentle, bear-shaped robot, helps lift and transfer patients with mobility challenges, reducing physical strain on human caregivers. This not only improves workflow efficiency in care facilities but also reduces the risk of caregiver injuries, enabling better long-term staff retention and quality of care. In the United States, Moxi is a service robot designed to assist healthcare staff by performing routine logistical tasks, such as delivering supplies or medications. By taking over routine, time-consuming chores, robots like Moxi allow human caregivers to focus on more engaging, relationship-centered aspects of care.

8.5 International Perspectives on AgeTech

International examples underscore the versatility and impact of AgeTech:

- Japan, facing one of the oldest populations in the world, is pioneering the use of robotic caregivers and AI-driven solutions to offset workforce shortages.
- Canada's AGE-WELL network has become a global leader in fostering AgeTech innovation, catalyzing partnerships across industry, academia, and community organizations.
- The United States, propelled by private sector leadership and responsive policy shifts, has expanded telemedicine coverage and integrated AgeTech into both clinical and home settings.
- European countries have long prioritized universal design principles, ensuring that assistive devices and inclusive environments support seniors' autonomy and dignity.

8.6 Challenges and Ethical Considerations in AgeTech

However, AgeTech's rise also brings challenges. Issues related to the digital divide, some older adults may lack the skills, access, or comfort to use new technologies, must be addressed through education and affordable access [2].

8.6.1 Ethical Concerns Include

- Data privacy.
- Automation's impact on human interaction.
- Equitable distribution of AgeTech resources.

Policymakers, industry leaders, researchers, and advocacy groups need to collaborate, ensuring that AgeTech solutions remain inclusive, user-centered, and ethically grounded [3].

8.7 Conclusion

The ascent of AgeTech reflects a transformative shift in how societies worldwide approach aging. Rather than viewing aging as a "problem" to be managed, AgeTech re-envisions it as a natural, enriching stage of life that can be supported, enhanced, and dignified through innovation. By blending technological sophistication with human-centered design and evidence-based strategies, AgeTech solutions empower older adults to remain active, engaged, and independent for longer. This shift is critical given the extraordinary demographic changes underway—and the enormous societal stakes involved.

From wearable health devices that proactively manage chronic conditions to telehealth platforms that extend the reach of healthcare professionals into seniors' homes, AgeTech tools are reshaping the care continuum. Cognitive health technologies offer hope in detecting and mitigating cognitive decline before it severely impacts quality of life, while assistive devices help older adults maintain mobility, hearing, and vision, key factors for social participation and autonomy.

Smart home solutions, from IoT-enabled appliances to fall detection systems and home-based robotics, redefine what it means to "age in place." Seniors can now live in environments specifically attuned to their needs, where risk is minimized, comfort is maximized, and connectivity to care networks is ensured. This is a far cry from traditional models of eldercare that often required institutionalization or placed a heavy burden on family caregivers.

The global embrace of AgeTech reveals that no single solution fits all contexts. Canada's AGE-WELL network excels at forging collaborative ecosystems that bridge research, industry, and community. Japan's deep investments in robotics and AI highlight a determined effort to overcome labor shortages and cultural expectations around eldercare. The United States shows how policy flexibility and market-driven innovation can catalyze rapid expansion of services like telehealth. Europe's emphasis on inclusive design and assistive devices ensures that older adults remain an integral part of society, rather than marginalized.

Yet, AgeTech's promise is accompanied by challenges and responsibilities. Bridging the digital divide is paramount. If older adults lack digital literacy or access to affordable technologies, they risk being left behind. Likewise, privacy concerns must not be overlooked. Highly sensitive health data flows through these new systems, demanding robust cybersecurity measures and transparent privacy policies. Ethical debates arise around the use of AI for care tasks and the potential dehumanization of relationships between caregivers and care recipients. Ensuring equitable access—so that low-income seniors do not miss out on lifesaving or life-enhancing technologies—is another pressing concern. Governments and organizations must provide subsidies, public funding, and policy support so that AgeTech can become an inclusive force for good.

Addressing these issues involves a whole-of-society approach. Policymakers can set frameworks for fair reimbursement and quality standards. Industry players must commit to user-centered design, affordability, and clear communication about product capabilities and limitations. Researchers and clinicians should engage in interdisciplinary collaborations, blending medical expertise with engineering, design, and social science perspectives. Advocacy groups and non-profits can represent older adults' voices in AgeTech development, ensuring that solutions meet real needs rather than perceived wants.

Looking to the future, the integration of AgeTech into mainstream healthcare and senior living is likely to deepen. Advancements in AI, machine learning, and robotics will produce more intuitive interfaces and smarter systems capable of anticipating user needs. Brain-computer interfaces, advanced prosthetics, and precision medicine tailored to genetic profiles may become commonplace. 3D-printed assistive devices, personalized nutrition plans guided by data analytics, and even virtual reality-based cognitive therapies could expand the AgeTech repertoire.

For societies to fully benefit from these developments, a sustained commitment to inclusivity, education, and participatory design is essential. Seniors themselves must be involved in technology testing and feedback loops, ensuring that products respond to their lived experiences. Communities should embrace intergenerational solidarity, where younger people assist older adults in navigating new tools, creating a culture of mutual support.

Ultimately, AgeTech is more than a set of gadgets or platforms. It represents a paradigm shift in care, community life, and social values. By recognizing older adults as active, integral members of society who deserve agency and respect, AgeTech and its stakeholders can help forge a future where aging is not an insurmountable challenge but an enriching phase of human life. The unprecedented demographic changes unfolding worldwide demand nothing less than this transformation. With thoughtful planning, ethical stewardship, robust innovation, and international collaboration, AgeTech can help ensure that our global communities embrace aging as an opportunity for growth, learning, and improved quality of life.

References

1. Beard, J. R., Officer, A., de Carvalho, I. A., Sadana, R., Pot, A. M., Michel, J.-P.,... Chatterji, S. (2016). The World report on ageing and health: a policy framework for healthy ageing. *The Lancet, 387*(10033), 2145–2154. https://doi.org/10.1016/S0140-6736(15)00516-4
2. Coughlin, J. F. (2020). The Fourth Wave of Technology and Aging: Policy Innovation to Ensure Equity and Inclusion. *Public Policy & Aging Report, 30*(4), 138–141. https://doi.org/10.1093/ppar/praa032
3. Lepore, M. J., & McMullen, T. (2024). Public Policy and Aging Research: After 50 Years of the National Institute on Aging. *Public Policy & Aging Report, 34*(4), 111–113. https://doi.org/10.1093/ppar/prae025
4. United Nations. (2024). World Population Prospects 2024. Retrieved from https://www.un.org/development/desa/pd/world-population-prospects-2024

Digital Inclusion for Seniors: Bridging the Digital Divide and Designing User-Friendly Technologies

9

9.1 Introduction

The digital revolution has reshaped modern life in countless ways, influencing how people connect with each other, engage with healthcare systems, conduct their finances, participate in the workforce, and access government services. From smartphones and tablets to telehealth platforms and online banking, digital tools have streamlined and enriched many facets of our daily routines. Yet, even as these technologies proliferate, a significant segment of the global population struggles to enjoy their full benefits. Older adults, in particular, often find themselves on the less connected side of this "digital divide" [2].

According to the International Telecommunication Union (ITU), older adults are the least likely demographic to use the Internet, with just 58% of individuals aged 55 and above having online access worldwide in 2020 [4]. This gap in connectivity is not merely a technical matter, it has profound social, economic, and health implications. Seniors who remain offline face greater risk of isolation, reduced access to telemedicine and other digital health services, and missed opportunities for civic engagement and lifelong learning [5].

The reasons behind this divide are multifaceted. Economic constraints, low digital literacy, physical limitations, and concerns about security and privacy all discourage seniors from embracing technology. In countries like Canada, which has invested in programs like "Connecting Families" to improve digital inclusion, rural and low-income seniors still encounter significant barriers to meaningful connectivity. Meanwhile, in many low- and middle-income countries, limited broadband infrastructure and high costs create especially challenging conditions for older adults striving to get online.

This essay delves into the complex landscape of the digital divide for seniors, exploring the challenges they face, the importance of digital inclusion, and strategies for designing user-friendly, accessible technologies. Drawing on examples from Canada, Japan, Europe,

A. Mihailidis, *The Age of AgeTech*, Synthesis Lectures on Technology and Health,
https://doi.org/10.1007/978-3-032-04886-8_9

the United States, and beyond, it highlights global best practices and forward-looking solutions. As societies continue to age and technology advances, closing the digital gap for seniors is not simply a matter of equity—it is integral to sustainable aging strategies that recognize older adults as vital contributors to modern life.

9.2 The Digital Divide: Challenges Faced by Seniors

9.2.1 Access to Technology

Access encompasses not just the availability of devices, such as smartphones, laptops, and tablets, but also the infrastructure that supports them, notably high-speed Internet connectivity. Many seniors live on fixed incomes, relying on pensions or limited retirement savings. The cost of purchasing new devices, upgrading outdated hardware, or subscribing to broadband services may feel prohibitive, particularly if technology does not seem immediately valuable or user-friendly [1].

Geographic disparities compound these issues. Rural and remote areas often lack reliable Internet connections. Even in a country like Canada, known for its high standard of living, rural seniors struggle with slow or unstable service, limiting their ability to engage with telehealth, online banking, or digital government services. Similar challenges arise worldwide. In parts of the United States, rural broadband remains a pressing concern, while in many low-income nations, Internet infrastructure is minimal, leaving older adults disconnected from global information networks and essential online resources.

9.2.2 Digital Literacy

For technology to be truly empowering, users must understand how to navigate digital environments confidently. However, many seniors grew up in an era where digital technology was nonexistent or rudimentary. They may never have had formal training in using computers, the Internet, or smartphones [2].

Research suggests that older adults may experience anxiety or skepticism about technology, worried that they might not be able to keep pace with rapid changes. Without patient, accessible training and continuous support, seniors can become discouraged, reinforcing a self-perpetuating cycle of digital exclusion.

9.2.3 Accessibility Challenges

As people age, they may experience sensory, motor, or cognitive changes that affect technology use. Reduced visual acuity can make reading small text on screens difficult, while complex interfaces may overwhelm individuals with diminished fine motor skills. Conditions like arthritis can render extended typing or swiping uncomfortable. Similarly, cognitive changes, whether mild memory issues or more pronounced forms of cognitive impairment, can hinder the ability to learn new interfaces or remember passwords.

9.2.4 Security and Privacy Concerns

Older adults are often targets of online scams and fraud, heightening their wariness. Phishing emails, dubious pop-ups, and unscrupulous "tech support" calls have eroded trust in digital platforms. Seniors may fear identity theft, hesitate to conduct online financial transactions, or feel uncomfortable sharing personal information to access telehealth services. Without clear guidance, user-friendly authentication methods, and robust educational initiatives on digital safety, many older adults will continue to opt out of the digital world [1].

9.3 The Importance of Digital Inclusion

Ensuring that seniors can access and confidently use digital tools is not simply a matter of equity; it has profound implications for their autonomy, health, social connectedness, and overall well-being. Digital inclusion ensures that older adults can participate fully in an increasingly digital society.

9.3.1 Promoting Independence

Independence is a cornerstone of healthy aging. Digital tools help seniors maintain autonomy, from online grocery shopping to scheduling virtual medical consultations without relying on family members [5]. In Canada, platforms like Maple allow seniors to easily access doctors and specialists from home, reducing mobility barriers and wait times. This model empowers older adults to manage chronic conditions, schedule follow-ups, and receive timely care without leaving their living rooms.

9.3.2 Enhancing Social Connectivity

Social isolation is a widespread issue among older adults, particularly those living alone or in long-term care facilities. Technology can help bridge these gaps, connecting seniors to family, friends, and support networks. In Canada, the Telus Health Companion platform fosters virtual social networks, enabling seniors to interact with loved ones and caregivers remotely. Through video calls, group chats, and online activities, technology can reduce loneliness, promote mental health, and help seniors stay engaged with their communities.

9.3.3 Facilitating Healthcare Access

As populations age, the burden of chronic diseases grows. Effective management of conditions like diabetes, heart disease, or arthritis often requires consistent monitoring and frequent communication with healthcare providers. Telemedicine platforms offer a solution that is both cost-effective and patient-centered. Japan's Dokodemo Doctor platform, for example, connects seniors in rural areas with healthcare professionals through mobile devices. Older adults can receive timely guidance and prescriptions, avoiding the stress and expense of long journeys to distant hospitals.

9.3.4 Economic Participation

Digital inclusion also opens doors to economic opportunities. Seniors who master online banking can maintain financial independence, paying bills, managing investments, and tracking their pensions with ease. Engaging in e-commerce, seniors can access specialized products or even sell crafts and services. Ride-sharing services like Uber and Lyft, which offer user-friendly interfaces and accessible design features, enable older adults to secure reliable transportation, helping them remain active participants in their communities. Some seniors even explore gig economy roles, such as tutoring, consulting, or freelance work, supported by digital platforms that connect them with potential clients.

9.4 Bridging the Digital Divide: Global Strategies

Addressing the digital divide requires coordinated efforts from governments, private companies, non-profit organizations, and local communities. By tackling affordability, digital literacy, accessibility, and security concerns head-on, it is possible to create an environment where seniors feel empowered to embrace technology.

9.4.1 Affordable Access Initiatives

Cost is a fundamental barrier to digital inclusion. Policies and programs that subsidize Internet access and devices can have a transformative effect. Canada's "Connecting Families" program offers low-income households affordable high-speed Internet and discounted devices [3]. Similarly, the National Broadband Network (NBN) initiative in

Australia aims to improve Internet infrastructure nationwide, including in rural regions, thereby benefiting older adults who previously had no reliable connectivity.

Private sector partnerships can also play a role. Telecommunications companies, libraries, and community centers can collaborate to provide free or low-cost Wi-Fi hotspots, loaner devices, and local tech hubs where seniors can familiarize themselves with digital tools in a supportive setting.

9.4.2 Digital Literacy Training

Building confidence and skills is essential. Structured digital literacy programs tailored to seniors can demystify the online world. Singapore's "Seniors Go Digital" program, for example, offers free workshops that teach older adults how to use smartphones, navigate government e-services, and stay safe online. In the U.K., the "Digital Champions Network" trains volunteers to provide one-on-one support, helping seniors learn at their own pace without feeling rushed or embarrassed.

The emphasis in these initiatives is often on practical, hands-on training. Rather than overwhelming learners with technical jargon, instructors can focus on everyday scenarios: sending emails to grandchildren, setting medication reminders, or looking up recipes. Building digital confidence through small, incremental victories can transform technology from a source of anxiety into a valuable ally.

9.4.3 Accessible Design Principles

User-centered design is critical. Creating interfaces that accommodate visual, auditory, motor, and cognitive changes associated with aging can make technology less intimidating and more useful. Larger fonts, high-contrast color schemes, and simplified menus can all reduce cognitive load. Specialized devices like the GrandPad tablet are engineered with seniors' needs in mind, featuring large buttons, intuitive navigation, and pre-set contacts, making communication and browsing far more accessible.

Voice-activated assistants like Amazon Alexa and Google Assistant further ease the burden on older adults who struggle with touchscreens or keyboards. By leveraging voice commands, seniors can check the weather, adjust the thermostat, or request information without complicated physical interactions. Similarly, adaptive features such as screen readers, text-to-speech functionalities, and custom accessibility settings can empower seniors with visual or hearing impairments to engage with digital platforms independently.

9.4.4 Addressing Security Concerns

Trust and safety are foundational to digital inclusion. Without confidence in online platforms, seniors may opt out entirely. Education is a powerful tool. Financial institutions like RBC in Canada have developed workshops and resources to educate seniors on identifying scams, verifying authentic websites, and using secure passwords. Simplified authentication methods, such as biometrics, can reduce the cognitive burden of remembering complex passwords and reassure older adults that their data is protected.

Public awareness campaigns, helplines dedicated to digital safety, and partnerships with law enforcement agencies can further bolster confidence. Additionally, user-friendly privacy settings and transparent data use policies reassure seniors that technology companies respect their rights and personal information.

9.5 Designing User-Friendly Technologies

Bridging the digital divide for seniors is not simply about teaching them to use existing devices and platforms, it also involves reimagining how technologies are designed from the ground up. Human-centered design, inclusive product testing, and adaptive interfaces can ensure that technologies genuinely meet older adults' needs.

9.5.1 Human-Centered Design

Human-centered design places the user's experience at the forefront. Rather than retrofitting complex products for older adults, developers and engineers can engage seniors from the earliest design stages. Co-creation workshops, focus groups, and iterative prototyping sessions allow seniors to voice their preferences and pain points. Canada's AGE-WELL, for instance, champions inclusive design projects, inviting seniors to collaborate with researchers and developers to ensure that devices and applications are not only technically sound but also resonant with their daily lives and values.

9.5.2 Inclusive Product Testing

A technology that passes laboratory tests with flying colors may still confound a 75-year-old user with limited vision or mobility. Including seniors in usability testing helps identify subtle issues that might otherwise go unnoticed. Apple's Accessibility Labs, for example, test products like the iPad with older users to refine features such as magnification tools, enlarged onscreen keyboards, voice navigation, and gesture-based shortcuts.

Inclusive testing ensures that when a product finally reaches the market, it will genuinely be user-friendly to a diverse population.

9.5.3 Adaptive Interfaces

Adaptive interfaces that adjust to individual users' needs represent a growing frontier in inclusive design. Smartphones like Samsung's Easy Mode simplify interfaces, using larger icons and a more minimalist layout. In the future, machine learning algorithms might continuously adapt to users' abilities and preferences, automatically increasing font sizes, recommending video tutorials, or reorganizing menu structures based on patterns in user behavior. This kind of dynamic customization can ensure that technology evolves alongside the user, reducing frustration and increasing comfort over time.

9.6 Global Examples of Digital Inclusion

Achieving digital inclusion is a shared global challenge. While contexts differ, countries around the world are experimenting with policies, programs, and technologies that can serve as models and inspiration for one another. Examining these examples reveals common themes—affordability, training, accessibility—and how different cultures and policy environments address them.

9.6.1 Canada

Canada's efforts to promote digital inclusion are multifaceted. The Digital Literacy Exchange Program provides funding to non-profit organizations that teach seniors digital skills, fostering confidence and resilience in navigating the online world. AGE-WELL, Canada's technology and aging network, works with industry, community groups, and academic institutions to develop AgeTech solutions, from telehealth platforms to smart home systems, ensuring that seniors benefit from cutting-edge research and innovation.

This collaborative approach, bridging public and private sectors, academia, and civil society, underscores Canada's commitment to inclusive digital transformation. Initiatives rooted in community engagement ensure that seniors' voices guide the development of new tools, making technology more responsive to their needs.

9.6.2 Japan

Japan, with one of the world's oldest populations, has emerged as a pioneer in integrating technology to support older adults. Recognizing that digital inclusion is not just a matter of plugging seniors into existing systems, Japan invests in tailored workshops and community programs. The Tech-Savvy Seniors Program introduces older adults to digital tools through hands-on sessions, while companies like NTT Docomo produce simplified smartphones featuring large buttons, streamlined menus, and accessible interfaces.

The Japanese approach also emphasizes robotics and AI, exploring how social robots might support older adults in communication, medication reminders, and cognitive exercises. While not all seniors will embrace advanced robotics, the diversity of Japan's initiatives ensures multiple pathways to digital inclusion.

9.6.3 Europe

European nations have long prioritized social equity and universal design principles, values that naturally extend to digital inclusion. Sweden's Digital Participation Program provides subsidized devices and Internet access to older citizens who might otherwise remain offline. Germany's Digital Seniors Initiative pairs older learners with younger volunteers, forging intergenerational bonds that make the learning process more enjoyable and sustainable.

In the Netherlands and Denmark, libraries and community centers serve as digital hubs where seniors can practice their skills in a welcoming, low-pressure environment. These models demonstrate that digital inclusion is not purely technical, social infrastructure and cultural attitudes play a crucial role in guiding seniors into the digital realm.

9.6.4 United States

The U.S. employs a range of public–private partnerships to advance digital inclusion among older adults. The AARP Foundation's Connect2Affect Initiative specifically addresses social isolation, offering digital tools and resources to help seniors connect with community services, health providers, and social groups. Similarly, digital literacy programs led by libraries, non-profits, and community colleges help seniors learn basic computer skills, navigate health apps, and explore online services confidently.

The country's expansive market also fosters innovation: technology giants, healthcare insurers, and telecommunication providers experiment with platforms that cater to seniors' needs. As a result, older adults can choose from a variety of learning opportunities, devices, and service models, aligning their digital engagement with personal preferences and lifestyles.

9.7 Future Directions

The landscape of digital inclusion is dynamic, evolving as new technologies emerge and societies reassess their priorities. Several trends suggest future directions that may further empower seniors in the digital world.

AI and Machine Learning: Advanced AI algorithms have the potential to provide personalized support. Systems could learn from a user's interactions, simplifying interfaces, offering helpful tutorials at just the right moment, or suggesting adaptive features to accommodate visual or motor changes over time. AI-driven chatbots might guide seniors through complex tasks, from filing taxes online to comparing health insurance plans.

5G and Beyond Connectivity: As 5G networks become widespread, telemedicine can become more robust, offering high-quality video consultations, real-time health monitoring, and swift data exchange. Seniors living in remote areas stand to benefit significantly as lag-free video streaming and advanced sensor integration become standard, further bridging the urban-rural divide.

Augmented Reality (AR) and Virtual Reality (VR): Emerging immersive technologies have potential therapeutic and educational applications for older adults. VR-based simulations could help seniors practice mobility exercises or cognitive training in a safe environment. AR overlays might assist with medication management, overlaying instructions on pill bottles as seniors view them through a device. These technologies, once refined and made more accessible, could redefine how older adults interact with their environments and maintain cognitive and physical function.

Interdisciplinary Collaborations: The future also calls for stronger interdisciplinary collaborations. Geriatricians, occupational therapists, software engineers, product designers, and social workers can work together to develop tech tools aligned with older adults' holistic needs. Policymakers can incentivize such collaborations through grants, awards, and frameworks that emphasize user-centered outcomes.

9.8 Conclusion

Digital inclusion for seniors is not a peripheral concern—it is central to ensuring that older adults remain active, engaged participants in modern life. As societies grapple with rapid demographic shifts, extending connectivity, digital literacy, accessible design, and security assurances to seniors can promote independence, combat isolation, improve health outcomes, and foster economic participation.

Global examples underscore that solutions are most effective when they address multiple facets of the issue. Canada's community-driven programs, Japan's simplified devices, Europe's intergenerational training models, and the U.S.'s public–private partnerships each illustrate valuable approaches to making digital engagement more equitable.

By learning from these examples and investing in human-centered design, affordability initiatives, and robust educational programs, we can close the digital gap for seniors.

As technology evolves, so too must our strategies for inclusion. Future-oriented solutions grounded in AI, VR, and 5G connectivity, along with sustained policy efforts and stakeholder collaboration, will ensure that no senior is left behind in the digital era. Ultimately, digital inclusion is about respecting and empowering older adults, recognizing that the promise of technology should enrich all stages of life, and that every generation deserves a meaningful role in shaping our shared digital future.

References

1. Anderson, M., & Perrin, A. (2017). Tech adoption climbs among older adults. Retrieved from https://www.pewresearch.org/internet/2017/05/17/tech-adoption-climbs-among-older-adults/
2. Charness, N., & Boot, W. R. (2009). Aging and Information Technology Use: Potential and Barriers. *Current Directions in Psychological Science, 18*(5), 253–258. https://doi.org/10.1111/j.1467-8721.2009.01647.x
3. Government of Canada. (2024). About the New Horizons for Seniors Program. Retrieved from https://www.canada.ca/en/employment-social-development/programs/new-horizons-seniors.html
4. ITU. (2024). Measuring digital development: Facts and Figures 2024. Retrieved from https://www.itu.int/en/ITU-D/Statistics/pages/facts/default.aspx
5. Seifert, A. (2020). The Digital Exclusion of Older Adults during the COVID-19 Pandemic. *Journal of Gerontological Social Work, 63*(6–7), 674–676. https://doi.org/10.1080/01634372.2020.1764687

AI and Robotics in Elder Care: Companions and Caregivers 10

10.1 Introduction

I have always been struck by the way we talk about aging: we frame it as a problem, a crisis, and a burden. The sheer number of older adults is growing, a "silver tsunami" some experts call it, as though the rising population of people over 60 is a natural disaster we must brace ourselves against. By 2050, more than 2 billion individuals will be aged 60 and older [1]. Many of them will live with chronic conditions, cognitive impairments, and mobility limitations. Many will require long-term care. And we, as societies, face both unprecedented opportunities and daunting challenges to ensure they age with dignity and comfort.

Yet, when I look around at how we have prepared for this demographic shift, I see a lot of neglect. Our long-term care facilities are often understaffed and underfunded. Burned-out caregivers struggle to meet increasing demands. Even the wealthiest nations grapple with workforce shortages in healthcare, skyrocketing costs, and inefficiencies that would be scandalous if we were talking about almost any other industry. It is as if we have collectively shrugged our shoulders and decided that the old and frail can simply wait in line. We pay lip service to respecting our elders, while simultaneously placing them in conditions and systems that are barely sustainable.

A. Mihailidis, *The Age of AgeTech*, Synthesis Lectures on Technology and Health,
https://doi.org/10.1007/978-3-032-04886-8_10

Into this environment steps a set of powerful new technologies: Artificial Intelligence (AI) and robotics. We are told these innovations hold the key to a grand solution: that they can ease human burdens, enhance eldercare services, and even provide companionship to socially isolated seniors. The sales pitch is compelling—who wouldn't want robots that lift bedridden elders without risking a caregiver's back, or AI algorithms that predict heart failure episodes before they occur? For anyone who has watched an older loved one slip through the cracks of an impersonal care system, these technologies sound like a godsend.

But let's not get too starry-eyed. While I personally find the idea of AI and robotic caregivers intriguing and potentially transformative, I also fear we are rushing headlong into a brave new world without truly asking ourselves the tough questions. Are we using these machines to fill a gap we have allowed to fester through deliberate underinvestment and lack of political will? Are we outsourcing the compassion and human intimacy that eldercare demands to a series of microchips, motors, and code because it's cheaper and more convenient than treating older adults as fully human? While it is exhilarating to think that AI and robotics can improve the independence and quality of life for older adults, it is also provocative, and sobering, to consider that these very same tools might enable us to continue sidestepping our moral obligation to care for our elders with genuine humanity.

This essay dives deep into the applications of AI and robotics in eldercare, examining their roles as both caregivers and companions. It paints a global picture, drawing examples from Canada, Japan, the United States, and Europe, revealing how different societies are embracing or struggling with these new tools. Along the way, I will be honest and personal in my commentary: I am both a fascinated observer and a skeptical critic. I want to celebrate these breakthroughs while also demanding that we question them relentlessly. Because the stakes are too high to do otherwise.

10.2 The Role of Artificial Intelligence in Elder Care

We have long known that healthcare must become more proactive and personalized. For older adults, especially those juggling multiple chronic conditions, conventional models of care, short appointments separated by long waits, rushed check-ups, and reactive treatments are simply inadequate. AI, it is said, can break us out of this rut by continuously analyzing data to prevent crises rather than waiting for them to happen.

10.2.1 AI-Powered Health Monitoring

Consider the promise of AI-driven health monitoring: wearable sensors tracking heart rate, blood pressure, glucose levels, and sleep patterns, all feeding streams of data to algorithms that can detect anomalies well before they manifest as emergencies. In Canada, companies like AlayaCare have harnessed AI to analyze data from wearable devices and home sensors, offering early warnings for heart failure or diabetes complications. In the United States, IBM Watson Health applies AI to patients' electronic medical records, crunching complex datasets to recommend personalized treatment plans.

These examples seem almost miraculous, yet, they also raise pressing questions. Yes, AI can reduce hospital readmissions and save money. Yes, it can alert caregivers or clinicians when an older adult's condition starts trending downward. But what happens to the relationship between the care recipient and their human providers when machines shoulder so much diagnostic responsibility? Do we risk disempowering seniors by drowning them in graphs and warnings they cannot interpret without expert help? AI can guide us toward preventive care, but it should also remind us that data is no substitute for empathic communication and that cutting-edge tools should complement, not replace, informed consent and shared decision-making.

10.2.2 Virtual Healthcare Assistants

AI-powered virtual assistants, from chatbots to voice-activated interfaces, already help younger generations manage everything from pizza deliveries to online banking. For older adults, these assistants could be game-changers, helping them navigate healthcare appointments, understand medication schedules, and find reliable health information. Canada's Maple platform integrates chatbots to help seniors book telehealth appointments or manage prescriptions. In the United Kingdom, the NHS uses AI-driven chatbots as symptom checkers that tailor healthcare advice for older users.

But let's be honest: relying on a chatbot for healthcare guidance can be disconcerting. Seniors often prefer speaking with a real nurse or pharmacist. The fact that cost and shortage of professionals might push us to adopt virtual assistants is not exactly a mark of societal progress. It is more like a desperate patch. This shouldn't diminish the value of these tools, especially when human resources are stretched thin, but we must remain vigilant. Virtual assistants must be transparent about their limitations, and their creators must ensure that older users receive guidance that is accurate, compassionate, and free of the subtle biases that can creep into AI models. If we do this right, these AI helpers can empower seniors to become more informed healthcare consumers. If we do it wrong, we will be throwing them into a labyrinth of automated menus and emotionless scripts, which may only increase confusion and frustration.

10.2.3 Cognitive Health Support

One of aging's cruelest aspects is the rise of cognitive decline, from mild memory lapses to full-blown dementia. AI has already begun to stake a claim in this domain, promising earlier detection, personalized cognitive training, and ongoing support. AGE-WELL's research in Canada explores AI-driven memory exercises and cognitive monitoring tools. Japan's Hitachi Brain Health Initiative analyzes speech patterns to identify the earliest indicators of cognitive decline.

I find this particularly compelling. We have spent decades, and billions of dollars, on pharmaceutical interventions that, in many cases, barely slow cognitive deterioration. If AI can help us detect changes early and recommend interventions, such as mental exercises, lifestyle adjustments, or even social activities, then older adults may be able to preserve their autonomy for longer. Yet, I am also unsettled by the idea that we're turning the most human aspects of ourselves, our memory, our speech, our personality, into data points for algorithms to parse. We must ensure that this move toward hyper-rational analysis does not erode the dignity of older adults. They are not merely brains to be optimized. They're people, rich with experiences and deserving of care that respects their humanity first, and treats their symptoms second.

10.3 The Role of Robotics in Elder Care

Robotics captivate the imagination in a way few technologies do. Who hasn't marveled at the idea of mechanical helpers that never tire, never complain, and can perform tasks that strain human muscles or patience? In eldercare, robotics promise to do heavy lifting, literally and figuratively. But I must emphasize: robots can be awe-inspiring, yet they also risk transforming intimate human care into a mechanical transaction if we are not extremely careful.

10.3.1 Robotic Caregivers

Japan has led the world in developing robotic caregivers, not surprising given its rapidly aging population and shrinking workforce. Robear, a bear-shaped robot, can lift and transfer patients, saving caregivers from back injuries and strain. RIBA (Robot for Interactive Body Assistance) does something similar, minimizing the risk of injuries. These machines address a critical challenge: caregiving is physically demanding, and without robust support, human caregivers risk burnout or injury.

Still, the spectacle of a friendly bear-like robot lifting grandma out of bed raises questions about what we consider "care." Care is not just the physical movement of a body from bed to chair. It is the gentle words, the reassurance, and the sense of being touched by a fellow human who cares. Replacing that with a robot might free caregivers' time for more personal tasks, but it might also reduce even those fleeting moments of empathy. We must resist the temptation to offload all the "dirty work" of caregiving onto machines, turning caregiving into a sanitized, mechanical process. Let's use robots judiciously, as aids that free human caregivers from drudgery so they can focus on building relationships, rather than as convenient excuses to cut staff and budgets even further.

10.3.2 Companion Robots

Social isolation is one of the silent killers of aging, eroding mental health, exacerbating chronic conditions, and robbing life of joy. Enter the companion robot, engineered to provide conversation, emotional support, and entertainment. Japan's PARO, a robotic seal, has found its way into eldercare facilities worldwide, offering therapeutic companionship to dementia patients. Israel's ElliQ combines AI and a robotic interface to engage seniors in conversation, remind them of appointments, and encourage social activities.

I have mixed feelings about this. On one hand, I have seen the heart-wrenching loneliness that many seniors endure. If a robotic seal can coax a smile from someone with dementia, is that not progress? If an AI-driven assistant can give a senior some semblance of interaction when human visitors are scarce, should we not welcome it? Perhaps. But I also worry that we will come to rely on these robots as a cheap substitute for human connection. Rather than fixing the societal failures that allow older adults to languish in isolation, we slap a technological Band-Aid over the wound. In doing so, we tacitly accept a world where robots provide comfort that humans no longer bother to give. That is a profoundly troubling trade-off. Let's be very clear: companion robots should augment, not replace, human relationships.

10.3.3 Service Robots

In many institutions, caregivers waste hours doing mundane tasks: delivering medication, fetching linens, cleaning rooms, and preparing meals. Service robots like the U.S.-developed Moxi take over these chores, allowing human staff to spend more time with patients. Canadian initiatives with smart lifting robots help caregivers perform physically demanding tasks more efficiently.

This is arguably where robots shine brightest, improving efficiency and allowing human caregivers to do what only they can do best, connect on a personal level. If deployed ethically, service robots can help rebalance workloads so that overstretched staff actually have time to talk with residents, hold someone's hand, or share a quiet laugh. Yet, I fear the corporate impulse is to see this automation as an opportunity to cut costs further, slashing human staffing under the guise of efficiency. We must remain vigilant, ensuring that service robots become tools of empowerment rather than a pretext for neglecting the human side of care.

10.4 Companions and Caregivers: A Dual Role

It's worth noting that we are at a tipping point: these technologies are not just mechanical servants; they are also positioned as emotional companions. This dual role, caregiver and companion, challenges fundamental notions of what it means to care for one another.

10.4.1 Companionship for Social Well-Being

Social isolation can devastate an older adult's mental and physical health. Companion robots like PARO or ElliQ step into the gap, interacting with seniors, stimulating conversation, and offering a comforting presence. In Germany, the Giraffe Robot facilitates video calls with distant family members, while Japan's Aibo robotic dog entertains seniors with lifelike behaviors.

These examples tug at my heartstrings: imagine an older person who has not heard a loving voice in days, lighting up at the antics of a robotic puppy. Yet, I remain uneasy. Should our solution not be to ensure that no older adult is so isolated that only a robot is available to cheer them up? Relying on robots to fill this void risks normalizing a situation we should find unacceptable. We must challenge ourselves: can we design communities and policies so rich in human connection that robots remain optional enrichments rather than essential lifelines?

10.4.2 Enhanced Caregiving Through Automation

Similarly, when AI and robotics handle routine, repetitive, or physically grueling tasks, caregivers are theoretically freed up to spend quality time with the elderly. Perhaps that is the best-case scenario: a hybrid model in which human caregivers focus on empathy, communication, and skilled tasks, while robots deal with the heavy lifting. For example, Canada's ROSA (Rehabilitation Service Assistant) can help seniors with therapeutic exercises, reducing the load on physical therapists. Denmark's Elderly Care Bots provide mobility support so human caregivers can do what humans do best: listen, encourage, and empathize.

The challenge is trust. Will institutions genuinely reinvest the time and money saved by robots into improving human interactions? Or will they treat the introduction of technology as a cost-saving measure, reducing staff and stripping care of its fundamental humanity? We must set quality-of-care standards that ensure robotics and AI enhance, rather than erode, the human dimension of eldercare.

10.5 International Perspectives

Every nation grapples with aging populations differently, influenced by cultural values, political priorities, and economic constraints. By examining Canada, Japan, the United States, and Europe, we see both common patterns and intriguing divergences.

10.5.1 Canada

Canada invests in research networks like AGE-WELL, which funds studies on cognitive health tools and smart home robotics. Canadian startups like Kinova Robotics create assistive robotic arms to help individuals with mobility impairments. The Canadian approach often stresses collaboration between industry, academia, and community organizations, focusing on inclusive design and evidence-based policymaking.

As a Canadian observer, I appreciate these efforts. But I also want Canada to push harder, asking not just how robots can help seniors but why so many seniors feel isolated or underserved in the first place. Policies must ensure that technological innovation pairs with robust social safety nets, equitable healthcare funding, and genuine person-centered care. If technology becomes a substitute for meaningful investments in eldercare staff and infrastructure, then we are failing our seniors.

10.5.2 Japan

Japan is ground zero for the aging trend. With one of the world's oldest populations, it faces acute caregiving shortages. The government's CareBot and Nursing Robot programs aim to integrate robots into the majority of eldercare facilities by 2030. Robots are not a novelty in Japan; they are seen as indispensable to maintaining a functional eldercare system.

While it is easy to admire Japan's forward-thinking approach, I also wonder if it reflects a societal capitulation to demographic realities rather than a desire to humanize care. Integrating robots might alleviate workforce pressures, but it also might lock in a future where personal attention becomes a luxury. Japan's cultural acceptance of robots may ease this transition, but the global community should watch closely: will Japan's robotic eldercare model become a blueprint for others, and what will that mean for the human aspect of caregiving worldwide?

10.5.3 United States

The U.S. approach often leans toward commercialization and innovation. With CMS (Centers for Medicare & Medicaid Services) approving some AI-assisted technologies for reimbursement, and DARPA's Robotics Challenge incentivizing breakthroughs, the U.S. fosters a competitive market-driven environment. This can accelerate innovation but also risks prioritizing profit over people. America's patchwork healthcare system often leaves the poorest and most vulnerable seniors behind, so introducing advanced robots and AI will not inherently solve issues of inequality, unless strict policies ensure these tools reach those who need them most, not just wealthy families who can afford upscale assisted living facilities.

10.5.4 Europe

European countries tend to emphasize inclusivity, sustainability, and ethical considerations. Denmark's Elderly Care Robotics Program focuses on aging in place, using robots to help seniors remain in their own homes. Sweden's Robot Companions Initiative examines cultural values and ethics, ensuring that robotic caregiving aligns with European sensibilities around dignity and social welfare.

This emphasis on ethics and inclusivity is commendable. Europeans often prioritize the human element: how can we ensure that technology supplements care without diminishing it? The European approach might be more measured, ensuring that robotics and AI are integrated thoughtfully, respecting older adults' preferences and forging a path that other regions might emulate.

10.6 Challenges and Ethical Considerations

The road to fully integrating AI and robotics into eldercare is paved with obstacles and moral dilemmas. We risk forging ahead without pausing to address these urgent questions.

10.6.1 Cost and Accessibility

Robots and advanced AI systems do not come cheap. While governments and private companies may subsidize costs in well-funded programs, what about low-income seniors? What about under-resourced nursing homes in rural areas? If these technologies remain exclusive luxuries, we risk deepening healthcare inequalities.

Public–private partnerships could alleviate this, but only if guided by moral clarity. We must demand that these innovations serve everyone, not just a privileged subset. Ethical policymaking should ensure that these tools do not become another way to ration quality care based on income or geography.

10.6.2 Data Privacy and Security

AI relies on sensitive health data, raising the specter of data breaches and unethical data mining. Seniors, often less tech-savvy, may be more vulnerable to manipulation. Transparent data governance frameworks and strict cybersecurity measures are essential. We must guarantee that personal health information won't be sold to marketers or insurers who might use it against older adults. The integrity and safety of this data must be sacrosanct.

10.6.3 Balancing Automation and Human Touch

This might be the core dilemma: can we automate eldercare without losing its soul? Machines excel at consistency and efficiency, but they lack human intuition, empathy, and genuine emotional presence. We must resist the fantasy that robots will replace human caregivers entirely. Instead, we must craft hybrid models where robots free humans from drudgery, not from caring. That is a moral imperative. If we lose sight of the fact that caregiving is inherently human, we risk reducing elders to tasks and metrics. That would be a betrayal of our most cherished values.

10.7 Future Directions

As AI and robotics advance, we stand at the cusp of even more transformative changes. Personalized AI assistants may one day learn our quirks and medical histories intimately, adapting recommendations in real time. Autonomous vehicles could give older adults back their freedom of mobility. Rehabilitation robots might refine physical and cognitive therapy to the point where seniors recover capabilities once thought forever lost.

But these futures are not guaranteed. They depend on the choices we make today. Will we regulate these technologies to ensure equitable access? Will we foster interdisciplinary collaboration, bringing ethicists, gerontologists, and older adults themselves into the design process? Will we demand that the profits generated by AI and robotics be reinvested into eldercare systems, rather than lining corporate pockets?

We must also consider the cultural shifts required. Perhaps we need a global conversation about aging, reframing it not as a problem but as a natural part of life that deserves celebration and support. If we view older adults as valued members of society, keepers of wisdom, stories, and perspective, then the technologies we develop will reflect that respect. If we see them as inconvenient burdens, technology will become an excuse to warehouse them with machines that do the bare minimum.

10.8 Conclusion

AI and robotics are redefining eldercare in ways that would have seemed like science fiction just a generation ago. From AI-powered health monitoring to robotic companions, these tools promise to solve workforce shortages, enhance caregiving efficiency, and support older adults in maintaining their independence and well-being. They offer hope in a world that often seems indifferent to the needs of its elders.

Yet hope must be tempered with honesty. These technologies do not exist in a vacuum. They arise in societies that often undervalue caregivers, underfund eldercare, and treat older adults as afterthoughts. If we are not careful, AI and robotics will become high-tech Band-Aids on a fundamentally flawed system. They could enable us to avoid confronting our failures, substituting mechanical comfort for real human presence, and perpetuating a status quo where compassion is in short supply.

We owe it to ourselves—and to future generations who will someday be old and vulnerable—to do better. We should embrace AI and robotics, not as cheap fixes, but as catalysts for reimagining eldercare. Let's demand that the savings from automation fund better training and pay for caregivers, improved facilities, and more personalized attention. Let's insist that seniors are consulted in the design of these tools, ensuring their dignity and preferences shape the outcome. Let's use technology to enhance human connection, not erode it.

The world is aging rapidly, and the stakes are immense. AI and robotics can help us navigate this uncharted territory, but only if we have the courage to ask the hard questions, the integrity to set high ethical standards, and the empathy to remember that care is, at its core, a human act. If we rise to that challenge, we might just create a future where growing old is not something to dread, but a phase of life supported by extraordinary innovations that reflect our deepest values. If we fail, we will be left with a world of immaculate robotics and sophisticated AI, cold comfort for those who need real compassion in their final years. The choice, ultimately, is ours.

Reference

1. United Nations. (2024). World Population Prospects 2024. Retrieved from https://www.un.org/development/desa/pd/world-population-prospects-2024

Ethical Considerations in AgeTech: Privacy, Security, and Balancing Automation with Human Touch

11

11.1 Introduction

I have spent years immersed in the unfolding drama of AgeTech, the flashy new label for an ever-growing suite of robotics, AI-driven health tools, and assistive gadgets aimed at older adults. What started as a small wave of "senior-friendly" devices and apps has now become a global phenomenon, popping up in everything from clinical trials to corporate boardrooms. To be perfectly blunt, it is about time our society realized that technology can do much more for older people than just sound an alarm when they fall. But as promising as these high-tech wonders might be, I am increasingly rattled by the ethical minefields they drag into the spotlight: specifically, issues of privacy, cybersecurity, and that elusive "human element" which robots will never truly replicate.

It is not enough to stand on the sidelines and marvel at what robotic caregivers and advanced health trackers can do. The data they collect, intimate details about health, behavior, and location, makes them ripe targets for exploitation if we are not careful. Meanwhile, we are at risk of turning over so much caregiving to machines that older adults might end up feeling like lab specimens under constant surveillance or chess pieces shuttled around by cold, unfeeling arms. In my view, the rapid adoption of AgeTech raises some pressing questions: Will we prioritize the autonomy and dignity of older adults, or will we treat them as mere data points in a grand experiment?

Yes, we stand at a critical juncture. I firmly believe AgeTech could revolutionize elder care, but only if it is done right, and "doing it right" means tackling privacy and security head-on; preserving real human touch; and confronting uncomfortable truths about autonomy, consent, and potential biases. Throughout this essay, I am going to put these uncomfortable questions front and center. We will examine international guidelines, consider how different countries are addressing these concerns, and highlight what I believe are the possible paths forward—paths that I hope remain grounded in decency and respect

A. Mihailidis, *The Age of AgeTech*, Synthesis Lectures on Technology and Health,
https://doi.org/10.1007/978-3-032-04886-8_11

for older adults. After all, if we cannot get these basics right, then what is the point of all this technological progress in the first place?

11.1.1 Privacy and Security in Health Data

Health data is a veritable gold mine, especially when it concerns older adults. On the one hand, real-time tracking of heart rate, blood pressure, and daily movement can help clinicians customize care like never before. On the other hand, such granular data, some of it incredibly personal, turns older adults into prime targets for identity thieves, snoopy corporations, and even manipulative family members. It's one thing to collect a daily step count; it is another to record every medication, bodily function, and perhaps even emotional state. If we don't get serious about locking down this information, we are not just breaching privacy, we are playing with people's lives.

11.1.2 The Importance of Health Data Privacy

I have seen more than a few so-called "innovations" that pay mere lip service to user privacy. Let's be clear: without trust, the entire AgeTech ecosystem will collapse under its own weight. No older adult should have to wonder whether their medical or personal data might be sold off to advertisers or funnelled into some nefarious black market.

In Canada, the Personal Information Protection and Electronic Documents Act (PIPEDA) has at least tried to lay down some guardrails by forcing organizations to justify why they collect data and how they use it. Projects under AGE-WELL also tout a privacy-first approach, claiming they weave ethical considerations right into the product's DNA. That is a good start, but I am still pressing for more explicit, user-friendly clarifications, especially since so many older adults are overwhelmed by digital jargon. If we want seniors to embrace telemedicine and wearable health monitors, we had better be crystal clear about how their information is protected.

For me, privacy is not just a legal formality, it is a moral obligation. Older adults will share data if they feel safe and respected. That trust cannot be earned by burying crucial details in a 20-page privacy statement that no one reads. We need to ensure that every system, whether it is a wearable ECG monitor or a smart home device, comes with easy-to-understand terms and a user experience that allows for genuine informed consent.

11.1.3 Cybersecurity Challenges

Now, let's talk about the elephant in the room: cybersecurity. A single data breach can have catastrophic ripple effects, exposing older adults to fraud or, worse, tampering with medical devices that keep them alive. It is mind-boggling how many IoT-enabled devices still come with subpar protection—some do not even require strong passwords. That is as good as laying down a welcome mat for hackers.

In the United States, the Food and Drug Administration (FDA) requires cybersecurity risk assessments, which forces companies like Medtronic to up their game by implementing encryption and multi-factor authentication. This is a step forward, but let's not kid ourselves: malicious actors are getting more sophisticated by the day. We need real enforcement, not just guidelines that can be skirted or watered down. Moreover, older adults often lack the know-how to maintain device security, making them low-hanging fruit for cybercriminals. Leaving this population to fend for themselves in a sea of phishing scams and suspicious links is simply irresponsible.

And let's not forget that internal threats can be just as dangerous as external ones. Healthcare providers or employees with legitimate access can easily misuse or leak information if adequate oversight is missing. There needs to be a culture of accountability built into every link of the chain, from device manufacturers to caregivers to the IT staff who manage the servers.

11.1.4 Informed Consent

I cannot overstate the importance of true informed consent. And by "true," I mean older adults actually knowing what they are signing up for, not just tapping "accept" on a mind-numbingly long user agreement. Many of the seniors I have worked with or spoken to have admitted they rarely read privacy disclosures. Even if they try, the language is often so convoluted that it might as well be in code. This is no different for the majority of us reading this book right now.

Singapore's Seniors Go Digital Program has the right idea: actually teaching older adults how to navigate the digital landscape, including how to spot and avoid privacy pitfalls. But it is not enough to say, "Hey, here's a program, go sign up!" We need systemic efforts to put these educational tools in front of older adults, whether through local community centers, senior advocacy groups, or even faith-based organizations. Unless the entire ecosystem steps up, older adults will remain easy prey for exploitative data practices.

11.1.5 On Data Security, Institutional Trust, and the Fragility of AgeTech Adoption

Despite the promise of AgeTech to transform aging in place, its long-term sustainability depends on a foundation that is becoming increasingly fragile: data privacy and institutional trust. As we entrust more of our lives, and those of vulnerable populations, to connected systems, we must face a sobering reality: no data system is completely secure. Even well-funded federal institutions with extensive firewalls have experienced breaches, and smaller companies developing AgeTech solutions face even greater risks. This is not just a technical issue but also a social and psychological one. For many older adults, trust in digital systems has been damaged by previous negative experiences, and no amount of user education can fully restore that trust without systemic changes.

Equally concerning is the lifecycle of data once collected. In a time of frequent mergers and acquisitions, it is crucial to consider what happens when a small startup, trusted by its users for its mission and transparency, is acquired by a larger conglomerate, possibly operating under different data governance laws. When cloud-based systems cross regulatory boundaries, personal information may be repurposed or exposed in ways that contradict the original intent. This global complexity underscores the need for stricter data residency requirements, informed consent protocols, and privacy-by-design strategies that are enforceable across jurisdictions.

There is also a deeper ethical dilemma at play. If the information generated by AgeTech tools can be used not just to support but to penalize, such as triggering institutional responses, insurance decisions, or the loss of housing, then the technology may inadvertently become an instrument of coercion rather than care. When older adults believe that activating a device might result in being removed from their homes, the utility of the tool is nullified. In such scenarios, surveillance becomes disempowering, and self-protection may look like silence.

To meaningfully address these challenges, AgeTech must be designed not only with the user in mind but also with a critical understanding of the policies, incentives, and systems into which it is embedded. Trust cannot be built on optimistic design alone; it must be undergirded by structural guarantees, legislative protections, and a culture that values elder autonomy over institutional convenience.

11.1.6 Solutions

The promise of AgeTech rests not only on technical innovation but on moral clarity. As devices enter the private spaces of older adults, listening, watching, learning, they also inherit a profound responsibility. That responsibility is not just to function or to delight, but to protect, to respect autonomy, and to earn trust in environments where that trust has been repeatedly broken.

Too often, the conversation around elder care technology celebrates progress while bypassing the darker realities of systemic neglect, institutional misuse, and corporate indifference. When data can be weaponized, by insurers, care providers, or even opportunistic relatives, the consequences are not theoretical. They are lived. And they are irreversible.

- **Transparent Policies**: No more legalese. We owe it to seniors to provide clear-cut, accessible explanations, ideally through short videos, large-print handouts, or narrated bullet-point summaries that clearly state what is being collected, who has access, and under what conditions. This is non-negotiable if we are serious about protecting them. Any device that operates in an elder's home should have a simple, on-device toggle that allows users to instantly review or disable certain streams of data collection without needing to call a support line or navigate a complex app. Consent must be ongoing, revocable, and easy.
- **Digital Literacy Programs**: Efforts like Singapore's "Seniors Go Digital" are a good start, but not nearly enough. We must expand digital literacy training to include cybersecurity drills, simulated phishing detection exercises, and scenario-based learning that allow older adults to see how data can be misused. These sessions should also include frank conversations about data brokerage, surveillance capitalism, and institutional overreach, especially in care contexts. When older adults are confident in their knowledge, they are far less likely to be manipulated or silenced by fear.
- **Data Sovereignty by Design**: Technologies deployed in elder care settings should adhere to data residency laws that keep personal information stored within the country of the user. If a startup gets acquired by a foreign conglomerate, there must be legal safeguards to prevent data from being repurposed, commodified, or transferred to jurisdictions with weaker protections. Devices that rely on cloud services must provide users with an option for local-only data processing, even if it comes with limited functionality. We should treat cross-border data flows as a critical design constraint, not an afterthought.
- **Ethical Firewalls Against Institutional Abuse**: We must openly confront the fact that AgeTech data may be used against its users, whether by insurers, care networks, or family members with legal power. If an AI system flags a user as "high risk," could that information be used to deny coverage, revoke independent living, or escalate to adult protective services without consent? These are not hypotheticals. Developers must build ethical firewalls into platforms: enforceable limits on how and when data can be accessed by third parties. Every system should include a clear "no-report" mode where elders can interact with tech privately unless they explicitly opt in.
- **Sunset and Buyout Clauses for Data**: Every AgeTech company should be required to define clear data-handling protocols in the event of acquisition, bankruptcy, or product discontinuation. This includes automatic deletion of user data if the company changes ownership, or clear re-consent requirements if data will be used under new terms. Aging users should not be subjected to the invisible churn of the tech economy. If a device becomes obsolete, the data it collected must not live on in a server farm halfway around the world.
- **Trust, Not Surveillance**: Finally, AgeTech must make a sharp distinction between care and control. Technologies that "listen" or "watch" must be reframed around consent, usefulness, and autonomy. If a person believes that falling will trigger institutional

removal, they will simply hide the fall or avoid using the device. The goal is not omniscience, it is earned trust. This will require reimagining AgeTech not just as a service layer but as a civic contract, where rights, dignity, and control reside squarely with the elder.

If we refuse to raise these standards, we are not just tolerating flaws, we are actively endorsing the exploitation of a vulnerable population. The urgency is real, and the moral calculus is straightforward: Either we design AgeTech to empower and protect, or we risk building systems that disempower, surveil, and punish. There is no neutral ground.

11.2 Balancing Automation with Human Touch

I will be blunt: the zeal for automating everything in elder care has gone off the rails. Robots designed to lift, feed, or even "interact" with older adults might be convenient, and they certainly relieve the strain on human caregivers. But let's not fool ourselves. What does it say about our society if we are more willing to invest in robotic arms than in strengthening the human workforce, that is, people who can offer empathy, conversation, and genuine warmth? While I am not anti-robot, I worry we are on the verge of turning caregiving into a transactional, mechanical process that neglects the deeply human aspect older adults crave.

11.2.1 The Role of Automation in Elder Care

No one can deny the efficiency gains. Japan's Robear robot, for instance, offers an impressive solution for physically demanding tasks, easing the stress on overworked nurses. But are we comfortable letting machines handle intimate caregiving tasks like bathing or toileting? Is it progress if we rob older adults of human contact at some of the most vulnerable moments in their daily lives? Efficiency is important, but we have got to be wary of crossing lines that turn aging into a dehumanizing process managed by software and steel.

Yes, automation can free up human caregivers for more meaningful interactions, but only if institutions use their new free time that way. Too often, the budgetary or staffing crunch in long-term care facilities leads to situations where robots do the heavy lifting and the humans are still overworked, leaving older adults in an emotional void. Automation is not a magical cure unless we fix the broader, systemic issues in how we value and support human caregiving.

11.2.2 The Emotional Dimension of Care

Let's state the obvious: older adults are human beings with emotional needs, memories, fears, and aspirations. No matter how advanced AI becomes, it cannot replicate genuine human empathy. Sure, a robot can simulate a smile or repeat comforting phrases, but it cannot truly understand or share in someone's emotional experience.

Sweden's Elderly Care Bots Initiative tries to strike a balance, using robots for mundane, repetitive tasks while ensuring humans handle emotional support. It sounds good on paper. But the question is: will these robots slowly creep into emotional territories that should remain human domains, all in the name of "efficiency"? I am sounding the alarm here because once we go down that path, once we normalize robotic companionship in place of real human interaction, we risk severing one of the most fundamental aspects of caregiving: authentic human connection.

11.2.3 Designing Human-Centered Automation

If automation must be part of elder care, and let's face it, it is already entrenched, we need to design it so that it complements, rather than engulfs, human caregivers. Canada's AlayaCare platform is an example of how AI can aid caregivers by managing logistics without marginalizing them. But I still see a world of difference between behind-the-scenes data analytics and a smiling robot that tries to replace your nurse's comforting hand.

We should involve older adults and caregivers in the design process from day one. Ask them how the technology can best serve their needs, not just how it can reduce costs or ramp up productivity. If we skip this integral step, we will end up with robots that feel alien, invasive, or downright condescending—an insult to those they are supposed to help.

11.2.4 Solutions

- **Companion Robots**: Tools like PARO, the robotic seal, can offer some soothing interactions, particularly for older adults with dementia. The key is never to promote these as a substitute for genuine human relationships.
- **AI-Assisted Tools for Caregivers**: When AI is used to handle paperwork or repetitive administrative tasks, it can enhance—rather than erode—the caregiver-patient relationship. This approach is far more ethical and sustainable than trying to turn robots into best friends for seniors.

In short, automation is not inherently evil, but how we deploy it reveals our priorities. We can champion efficiency while still reminding ourselves that human dignity and connection are not negotiable.

11.3 Ethical Implications of Assistive Technologies

Assistive technologies like GPS trackers and smart home systems can be game-changers. They can help prevent dangerous wandering in dementia patients or send automatic alerts if someone falls. Yet, these wonders of modern tech are not merely neutral tools; they carry the weight of ethical trade-offs, especially around autonomy and consent. When you install a system to monitor every move or track every breath, you are treading a fine line between safeguarding health and violating personal freedom.

11.3.1 Autonomy Versus Control

I have encountered many families who wrestle with how much to monitor a loved one with cognitive decline. Naturally, they want to prevent accidents, but constantly tracking someone can make them feel imprisoned. Are we truly okay with turning older adults into data streams? Sure, it might keep them physically safer, but at what cost to their emotional and psychological well-being?

The Alzheimer's Society in the UK advises using monitoring technologies with a hefty dose of caution, emphasizing the need for consent wherever possible. I applaud these guidelines. The big question, though, is whether we are going to follow them or twist them into convenient loopholes. If we are slapping trackers on older adults without their permission or knowledge "for their own good," we are trampling on their dignity in the process.

11.3.2 Bias and Accessibility

I cannot stress enough how easy it is for biased algorithms or non-inclusive designs to wreak havoc. If speech-recognition tools cannot parse certain accents or dialects, they become useless for large swaths of the older population. And if an AI system is trained primarily on data from urban hospitals, it might fail rural users who have different care requirements. The result? Entire groups of older adults shut out from the supposed "benefits" of AgeTech.

Accessibility also goes hand in hand with affordability. Fancy assistive devices might dazzle the market, but if they come with a hefty price tag or require broadband Internet that rural seniors do not have, we are just deepening the gulf between the haves and have-nots.

11.3.3 Solutions

- **Inclusive Design**: Why is it so revolutionary to involve older adults in focus groups and beta testing from day one? It should not be. Diverse participation ensures we're not unintentionally building a wall around certain demographics.
- **Bias Audits**: Every AI system that claims to be revolutionary in elder care ought to undergo rigorous, periodic audits. Enough with secret black-box algorithms; we need transparency to identify bias and fix it at the root.

Making assistive tech truly inclusive and unbiased is not some lofty ideal; it is a moral requirement if we do not want to leave large segments of the older population in the dust.

11.4 International Frameworks for Ethical AgeTech

It is tempting to think of ethics as an "optional extra." Yet, if we look around, many governments and international bodies are scrambling to catch up with the AgeTech wave, each one drafting guidelines, regulations, and best-practice recommendations. While the specifics vary, the underlying message is clear: older adults deserve technology that does not exploit them, rob them of privacy, or treat them as guinea pigs.

11.4.1 Canada

I have been impressed by Canada's AGE-WELL network (yes, I may be biased on this one considering I was the inaugural Scientific Director of this network), which insists on weaving ethics directly into the research and development process. They require projects to address equity, privacy, and inclusion head-on. Coupled with PIPEDA, which outlines how personal data can be used, Canada has set a relatively high bar for respecting older adults' rights. But let's not pat ourselves on the back too quickly. True enforcement and robust public understanding of these frameworks are essential if they are to have any teeth.

11.4.2 Japan

No country has a starker demographic time bomb than Japan, and it's responded with some of the most advanced robotics for elder care. Their Ministry of Health, Labour and Welfare demands that any such robotics put user safety and comfort first, which aligns with Japan's cultural emphasis on respect for elders. Still, I worry that economic pressures to fill caregiver shortages could overshadow the cultural ethos. When push comes to shove, will older adults be forced to accept robotic solutions whether they want them or not?

11.4.3 European Union

The EU's GDPR is often hailed as the global gold standard for data privacy. By forcing companies to embrace "privacy by design," the EU has inadvertently boosted the ethical bar for AgeTech, too. If you are launching an AgeTech device in Europe, you cannot just sweep privacy under the rug. You have to prove data minimization, consent, and robust security. But even GDPR can feel toothless if governments do not enforce it rigorously. I say more audits, more fines, and more public accountability, so we don't end up with a regulation that is all bark and no bite.

11.4.4 United States

The FDA's Digital Health Innovation Action Plan at least tries to highlight safety and transparency in digital health tools, including AgeTech. Yet, the U.S. lacks a unified privacy law like GDPR, creating a patchwork of state and federal rules. This mosaic approach leaves a lot of gaps through which unscrupulous actors can slip. With the rapid proliferation of telehealth and wearable devices, particularly accelerated by the pandemic, it is high time the U.S. got its act together and established consistent, enforceable protections for older adults.

Together, these frameworks show that we can't just let the tech sector go unchecked. It's positive to see countries making efforts, but until these regulations have real enforcement, companies with deep pockets might find it cheaper to pay fines (when caught) than to invest in ethical compliance from the start.

11.5 Future Directions and Recommendations

So, where do we go from here? Do we let AgeTech keep racing ahead, enthralled by its shiny new gadgets, or do we pause and hammer out the ethical scaffolding we so desperately need? The path forward hinges on a collective effort to strengthen governance, expand accessibility, and design for genuine human connection—three pillars that, in my view, are non-negotiable.

11.5.1 Strengthening Ethical Governance

Let's take global standards seriously, applying them across borders. If we can ensure uniform ethical guidelines, spanning privacy protocols, caregiver responsibilities, and user safety, we won't have a fractured landscape where companies can sidestep stricter laws by operating in lenient regions. Independent ethics committees can offer checks and balances, evaluating projects from the earliest design phases. This should not be optional; it should be mandatory for any entity receiving public funds or aiming to bring products to a broad market.

Such committees must be diverse: ethicists, tech developers, geriatric specialists, and, crucially, older adults themselves. If we fail to include the very people these technologies are supposed to help, we will keep producing tone-deaf products that do more harm than good.

11.5.2 Enhancing Accessibility

Accessibility is more than just making devices "user-friendly." It is about ensuring that no one is priced out, digitally illiterate, or physically incapable of using these innovations. If we genuinely care about older adults, we cannot make essential devices a luxury. Governments should consider subsidizing core assistive technologies, especially for seniors with limited income.

- **Subsidized Programs**: Think of it as an investment. By reducing healthcare costs in the long run, governments can save a fortune and, more importantly, provide older adults with the tools to live independently and safely.
- **Training and Support**: It is high time we approached digital literacy as a fundamental right, not a perk. Every senior center, library, or community hub could serve as a resource for older adults to learn, practice, and troubleshoot with new technologies. Otherwise, those who most need these solutions will remain on the sidelines.

11.5.3 Designing for Empathy

The day we forget empathy in technology design is the day we lose the plot entirely. Yes, robots can do a million things, but they cannot replace a heartfelt conversation or a reassuring human presence. If AgeTech aims to provide care, it should reflect the full spectrum of human needs—emotional, social, and psychological—rather than just ticking boxes for efficiency.

- **Hybrid Care Models**: Let robots tackle physically demanding tasks or repetitive routines, but leave room for actual caregivers to form relationships and offer emotional support. This synergy is possible only if healthcare systems invest in both people and technology, rather than using robots as a cost-cutting measure.
- **Adaptive AI Systems**: Developers must go beyond clinical metrics; they should incorporate an understanding of emotional states and personal histories. If an AI can detect loneliness or anxiety, it should prompt a human intervention, not dole out a scripted line of fake empathy.

We should treat older adults as co-creators, not passive recipients. Only then can we hope to build solutions that uplift rather than sideline the human aspects of aging.

11.6 Conclusion

Let me be perfectly candid: AgeTech is either going to elevate how we care for older adults or morph into yet another industry that exploits the vulnerable for profit. There is little middle ground here. With each new AI-driven platform or robotic caregiver, we are faced with moral choices about privacy, security, and the delicate dance between machine efficiency and genuine human warmth.

And let's not pretend these are purely academic questions. They have real-world stakes: data breaches that ruin lives, caregivers replaced by machines that cannot provide a comforting hand, and biases that shut out whole demographics of older adults. Meanwhile, international frameworks—from Canada's PIPEDA to the EU's GDPR—attempt to lay down ethical lines in the sand. Yet, the effectiveness of any regulation hinges on our collective will to enforce it and demand accountability.

Moving forward, the path is clear to me: embrace technology but never forget the humanity that underpins genuine care. Foster a global environment where strong ethical governance is the standard, not the exception. Make accessibility a priority, because if a life-changing device is out of reach for most, it is not really life changing at all. And, above all, design for empathy, ensuring that the oldest members of our communities are not relegated to living in a glorified science experiment.

Let's refuse to treat ethics like a pesky footnote. Let's aim higher, building an AgeTech future that genuinely respects privacy, fortifies security, and, most importantly, never loses sight of the human heart beating behind every piece of data.

12 Is Scientific Peer Review Killing Innovation in AgeTech Solutions for Older Adults?

12.1 Introduction

Societies across the globe are experiencing a dramatic demographic shift, driven by increasing life expectancy and declining birth rates. According to the United Nations [3], the population of individuals aged 60 and older is projected to rise to 2.1 billion by 2050, signifying a profound transformation in the age structure of our world. This unprecedented expansion of the older demographic brings forth a range of both challenges and opportunities. Healthcare systems will need to adapt to an increased prevalence of chronic and age-related conditions, while social welfare and economic policies must be revised to accommodate longer lifespans and potential strain on pension systems. Beyond these challenges, however, lie significant opportunities: older adults represent a wealth of experience, knowledge, and potential contributions to society, and new technologies can help them maintain their autonomy and enhance their quality of life.

Within this context, AgeTech, encompassing an array of technological solutions designed to address the needs of older adults, has emerged as a rapidly evolving field. Often conceptualized as technology-driven tools or systems that support independent living and improved care for older populations, AgeTech covers a broad spectrum. It ranges from wearable health monitors and telehealth applications to robotic caregivers and cognitive support tools. These innovations have shown considerable promise in empowering older adults to remain in their homes longer, reduce reliance on in-person medical consultations, and potentially mitigate caregiver burden. Moreover, AgeTech can help bridge gaps in healthcare access, particularly in remote or underserved regions, by enabling telemedicine and remote monitoring services.

Despite the remarkable potential of AgeTech, numerous hurdles stand in the way of translating innovative ideas into practical, widely adopted solutions. Traditional frameworks of scientific validation and clinical research, while essential for ensuring safety

A. Mihailidis, *The Age of AgeTech*, Synthesis Lectures on Technology and Health,
https://doi.org/10.1007/978-3-032-04886-8_12

and efficacy, often impose time-consuming processes on developers. Lengthy clinical trials can extend over several years, and accumulating robust data to prove efficacy may be particularly challenging with emerging technologies. Regulatory mechanisms, likewise, can be slow to adapt to new categories of devices, especially when artificial intelligence (AI) or machine learning approaches are involved. Companies, especially startups and small enterprises, may lack the capital to endure such protracted timelines. As a result, many promising tools risk becoming obsolete before they ever reach the market.

An equally pressing issue stems from ethical and legal complexities. Since older adults, especially those with cognitive impairments, may be considered a vulnerable population, strict guidelines govern research involving them. Ensuring informed consent can be more complicated, and additional oversight is necessary to safeguard participant well-being. These precautions, while critical, can add layers of complexity to research protocols, further slowing innovation. At the same time, concerns about data privacy and potential misuse of personal health information loom large, particularly as more sophisticated and networked devices collect sensitive data.

AgeTech innovators, therefore, face a delicate balance. They must navigate the imperative to demonstrate scientific rigor through peer-reviewed research while striving to bring products to market quickly enough to meet urgent needs in eldercare. The risk is that by the time a technology secures all necessary approvals, user requirements might have shifted, or more advanced tools may have superseded earlier prototypes. Startups are especially vulnerable, as they operate with limited budgets and cannot always afford the costs and time required for multi-year trials.

The primary objective of this essay is to examine how the scientific peer review process, in conjunction with regulatory and ethical factors, can inadvertently hinder AgeTech innovation. It delves into the current landscape of these technologies, highlighting specific examples like robotic caregivers, smart home systems, and telehealth platforms. The discussion then explores the principal barriers, ranging from lengthy validation processes to ethical and resource constraints, that prevent AgeTech solutions from reaching their full potential. The concept of "island of misfit solutions" will be introduced to illustrate how promising innovations sometimes become stranded in developmental limbo, never achieving the adoption they merit.

Following this overview, the essay proposes a series of reforms aimed at accelerating AgeTech deployment without compromising safety or integrity. These reforms include adaptive regulatory frameworks capable of flexible oversight, alternative research methodologies that can streamline validation, collaborative public–private models to pool resources, and the adoption of ethical guidelines tailored specifically for AgeTech. A central consideration throughout is how to strike a balance: on one hand, there is an urgent need to expedite innovation to address the swelling demands of an aging population; on the other hand, we must preserve the ethical and scientific standards that protect vulnerable individuals and sustain public trust in new interventions.

Finally, the essay surveys global perspectives, featuring initiatives in Canada, Japan, the United States, and Europe, to illustrate how different policy and regulatory environments can either facilitate or impede AgeTech growth. By examining these diverse contexts, we gain insight into successful strategies and identify areas where greater harmonization or mutual learning could benefit all stakeholders. Ultimately, this essay seeks to demonstrate that, while current validation mechanisms are critical for preserving safety and efficacy, they can be refined in ways that are conducive to timely innovation. By embracing reforms that account for the unique nature of AgeTech, we can better ensure that technology advances serve their intended beneficiaries—older adults—promptly, ethically, and effectively.

12.2 The Current Landscape of AgeTech

AgeTech is a broad and multifaceted domain that encompasses various technologies specifically designed or adapted for older adults. Although different categorizations exist, three prominent areas are robotic companions, smart home technologies, and telehealth platforms. Each addresses distinct aspects of eldercare, from emotional well-being and daily activity monitoring to remote healthcare consultations. By exploring these areas in greater depth, we can appreciate both the promise of AgeTech and the complexities involved in bringing these technologies to market.

12.2.1 Robotic Companions

Robotic companions represent one of the most intriguing, and sometimes controversial, facets of AgeTech. Devices like PARO, a robotic seal developed for dementia care, stand at the intersection of social engagement and therapeutic intervention. PARO is designed to react to touch and sound, employing sensors to detect when it is being stroked or spoken to, thereby simulating behaviors of real animals. The central premise is that these robotic companions can alleviate loneliness, reduce stress, and potentially mitigate some behavioral and psychological symptoms of dementia.

However, the enthusiasm surrounding robotic companions is tempered by ethical debates. For instance, some critics argue that providing older adults, particularly those experiencing cognitive decline, with robotic devices that mimic companionship may constitute a form of deception. Questions arise about whether it is ethically permissible to use robots as a substitute for human contact, even if many older adults and caregivers find them beneficial. Additionally, robust, long-term efficacy studies are still limited. While short-term research shows promise in improving mood and social interaction, there is a paucity of large-scale, longitudinal data to confirm sustained benefits. This dearth of extended follow-up research reflects one of the broad dilemmas in AgeTech: balancing rapid deployment of helpful tools with the rigorous evidence required to validate them.

Another challenge is cost. PARO's price tag, which can hover around USD $6,000, places it out of reach for many individuals, care homes, or health systems with constrained budgets. Although various funding schemes and grants might subsidize the costs, high expenditures inevitably limit accessibility, particularly in low-resource settings. Consequently, the full potential of robotic companions to support older adults is seldom realized. Developers must reconcile these cost constraints with research expenditures required to garner the approvals that make large-scale adoption feasible.

12.2.2 Smart Home Technologies

Smart home systems are another significant segment of the AgeTech market, leveraging the Internet of Things (IoT) to monitor and assist with daily tasks. From motion sensors that track an older adult's movements to automated lighting and temperature controls, these technologies are designed to maintain a safe living environment. They can detect anomalies, such as a sudden lack of movement, that might indicate a fall or a medical emergency, and then alert caregivers or emergency services. This early intervention capability holds great potential for reducing hospital admissions and long-term care placements.

In Canada, for instance, the AGE-WELL network has piloted prototypes that track mobility patterns and detect deviations that could signal fall risk. By using machine learning algorithms, these systems can adapt to individual behavior and offer personalized alerts. Such real-time data not only reassures family members but also allows health professionals to make more informed decisions. The ultimate goal is to ensure that older adults can age in place, retaining independence while mitigating the risks that arise from living alone.

Despite their promise, integrating multiple IoT devices into a coherent and user-friendly interface poses technological and usability hurdles. Older adults may be deterred by complex setups and may require simplified user interfaces or robust training to benefit fully. Additionally, data privacy questions loom large since these sensors and devices gather continuous streams of information, much of it sensitive in nature. Ensuring compliance with data protection regulations, such as the European Union's General Data Protection Regulation (GDPR), is essential. Meeting these privacy standards can be expensive and technically demanding, particularly for small startups with limited resources.

12.2.3 Telehealth Platforms

Telehealth, which facilitates healthcare delivery via digital platforms, has gained considerable traction in recent years, a trend accelerated by the COVID-19 pandemic. For older adults with limited mobility or who reside in remote areas, telehealth can bridge critical gaps in access to healthcare services. Platforms like Canada's Maple or Japan's Dokodemo Doctor enable virtual consultations with physicians, mental health professionals, and other healthcare providers, sometimes within minutes of making a request.

Such real-time access reduces travel constraints, lowers the risk of infection, and offers immediate medical guidance. These advantages make telehealth particularly appealing in geriatric care, where timely intervention can avert complications arising from neglected health issues. However, despite its clear benefits, telehealth faces its own set of validation and regulatory challenges. Healthcare systems typically require evidence-based proof of a platform's efficacy for diagnosing and managing conditions. Gathering this evidence frequently involves randomized controlled trials (RCTs) or at least robust observational data. Telehealth solutions also must comply with telemedicine regulations, which can differ significantly across jurisdictions.

For startups eager to scale, these varying requirements across countries create additional complexity. A platform might gain approval in one region, only to find itself facing an entirely new set of standards elsewhere. Additionally, the digital divide, where not all older adults have reliable Internet access or the technological literacy to utilize telehealth services, further complicates widespread adoption. Thus, telehealth emerges as both an exciting frontier and a complex landscape that requires careful navigation of clinical, regulatory, and practical concerns.

Overall, the current AgeTech ecosystem brims with innovation, yet each promising category, robotic companions, smart home technologies, and telehealth platforms, underscores the intricate interplay between potential benefits, ethical considerations, and the rigors of scientific validation. This interplay sets the stage for the next section, which delves more deeply into the specific barriers to AgeTech innovation and how they manifest in real-world scenarios.

12.3 Barriers to AgeTech Innovation

Despite the evident promise AgeTech holds, a range of obstacles hamper the swift development, validation, and adoption of innovative solutions. These barriers are multifaceted, often involving scientific, financial, regulatory, and ethical dimensions. Four primary issues are especially salient: the protracted nature of validation processes, resource constraints faced by innovators, regulatory hurdles that struggle to keep pace with technological evolution, and the ethical and legal complexities intrinsic to research with older adults. By dissecting these barriers, we can better understand why so many promising AgeTech solutions fail to reach the market or remain in prolonged developmental limbo.

12.3.1 Lengthy Validation Processes

Scientific validation remains a cornerstone of healthcare-related technology, and rightly so, as patient safety and reliable efficacy must be paramount. However, when it comes to AgeTech, conventional validation methods, such as randomized controlled trials, can stretch on for multiple years. While these multi-year trials are necessary in certain contexts, particularly where invasive procedures or pharmaceutical interventions are involved, they may not be the most appropriate or efficient for every form of AgeTech. For example, AI-powered cognitive assessment tools that monitor speech patterns or track ocular movements to detect early signs of dementia must be vetted for their accuracy. Yet, the data needed often involve longitudinal observations of cognitive changes over time. Gathering such data can delay market entry by several years, during which the underlying technology may already have advanced.

Another contributing factor is that emerging technologies frequently require novel metrics for evaluating efficacy. Traditional outcome measures might not capture the nuanced benefits of certain digital interventions. This misalignment amplifies the time and effort needed to establish new protocols or adapt existing ones. As technology iterates rapidly, especially AI-based solutions, which can evolve almost continuously, trying to synchronize R&D cycles with conventional validation timelines becomes daunting. The result is that some prototypes become obsolete by the time they are rigorously vetted, thereby stifling the incentive for further innovation in the short term.

12.3.2 Resource Constraints

Resource limitations are a pervasive challenge in the AgeTech sector, most acutely felt by startups and small companies. While big tech corporations or established healthcare firms can funnel substantial funding into R&D, smaller entities often struggle to compete. Conducting comprehensive research, including pilot testing and clinical trials, demands a pool of expertise (clinicians, data scientists, user experience designers) as well as costly infrastructure (labs, specialized equipment, data management systems). Recruiting participants, especially older adults with specific conditions, adds another layer of complexity and expense.

For instance, even if a startup develops an AI-powered tool for detecting early dementia, it may lack the financial means to recruit hundreds or thousands of participants for validation. Without evidence from such large-scale studies, regulatory bodies may be hesitant to grant market approval. Consequently, many of these small innovators either dissolve or pivot away from the AgeTech space, limiting the pipeline of novel solutions. Programs like Canada's AGE-WELL exemplify collective efforts to support startups by offering funding, mentorship, and collaboration opportunities, but demand often exceeds available resources.

Moreover, the competitive marketplace of digital health amplifies these challenges. Larger companies can rapidly prototype, test in-house, and scale. They also have established channels for distribution and are better positioned to absorb the financial risks of a lengthy regulatory review. Smaller players, in contrast, risk running out of capital before they can secure any meaningful revenue, particularly if external funding is tied to tangible proof of concept and regulatory milestones. The resulting dynamic stifles diversity in the AgeTech space, as many emerging voices are effectively crowded out.

12.3.3 Regulatory Hurdles

Regulatory oversight is designed to ensure that new technologies, especially those that influence medical decision-making, do not compromise patient safety or privacy. Yet, these regulations are often based on paradigms aligned with traditional medical devices or pharmaceuticals. Consequently, they may struggle to accommodate evolving innovations like AI-based tools, wearable monitors, or robotic companions. Defining whether an AI-based cognitive assessment tool is a "medical device" or a "general wellness product" is not trivial and can significantly alter the approval pathway.

In the United States, for instance, the Food and Drug Administration (FDA) has been progressively updating guidelines to accommodate AI and machine learning. However, the path toward compliance can still be lengthy. Startups might spend months or years tailoring their products to align with the latest interpretations of existing rules, only to discover that new guidelines or clarifications have been issued, necessitating further changes. Inconsistent or rapidly changing regulatory environments lead to uncertainty for developers and investors, potentially deterring investment in AgeTech ventures.

Globally, the situation is even more fragmented. Each country or region may have its own rules for telehealth, data privacy, and medical device classification. Navigating this mosaic is resource-intensive, especially for smaller companies that cannot afford specialized legal counsel for each market. As a result, many promising technologies remain confined to a single region or never launch, depriving older adults elsewhere of potential benefits.

12.3.4 Ethical and Legal Complexities

Research involving older adults is subject to strict ethical considerations, and the stakes are even higher when cognitive impairment is a factor. Informed consent protocols must be robust, and researchers must be particularly careful to avoid any form of exploitation or undue pressure. Additionally, older adults with dementia or other cognitive challenges may require proxies or legal guardians to provide consent. All these measures, while ethically indispensable, add layers of complexity that can elongate research timelines.

The debate around robots like PARO also exemplifies broader ethical quandaries. Some argue that by presenting a machine as a companion, we risk deceiving vulnerable individuals, potentially compromising their dignity. Consequently, institutional review boards (IRBs) may impose more rigorous reviews or require additional safeguards. These measures can be costly, especially for smaller AgeTech developers. Similar complexities arise with AI-driven assessments that must handle large volumes of personal health data. Ensuring compliance with data protection laws, such as the GDPR in the EU, requires robust encryption and stringent data-handling protocols, which can be cost-prohibitive for companies operating on modest budgets.

Taken together, these barriers, lengthy validation processes, resource constraints, regulatory hurdles, and ethical complexities, construct a significant blockade in the AgeTech innovation pipeline. They are not insurmountable but addressing them requires thoughtful reforms that balance patient safety, data integrity, and the genuine need for rapid innovation. The next section introduces the concept of the "Island of Misfit Solutions" to illustrate the tragic consequences of these obstacles: many potentially groundbreaking tools end up abandoned or languishing in extended testing phases, never realizing their intended impact.

12.4 The Island of Misfit Solutions

In the domain of children's stories, the "Island of Misfit Toys" is a place where unique but misunderstood playthings drift into obscurity. By analogy, the "Island of Misfit Solutions" describes a reality in which promising AgeTech innovations become stranded in developmental limbo. They are neither fully validated for widespread clinical or commercial deployment nor entirely discarded; rather, they exist in a protracted state of half-realization. This phenomenon results from the confluence of regulatory, financial, and ethical complexities previously outlined, and it ensures that many older adults remain unaware of, or unable to access, solutions that could significantly improve their daily lives.

12.4.1 AI-Powered Cognitive Tools

AI-driven tools for early detection of dementia stand as a poignant example of solutions that can become stranded. These systems often rely on analyzing large sets of data, ranging from speech patterns and language usage to ocular metrics or subtle motor function, to identify deviations that may signify cognitive decline [1]. In theory, early detection allows for prompt interventions that could delay or mitigate disease progression. Yet, in practice, bringing such a tool to market is fraught with obstacles.

First, developers must determine whether their technology qualifies as a medical device or a general wellness product. The distinction is critical; medical device classification triggers stringent regulatory requirements involving clinical trials and specific documentation. Second, the validation process requires longitudinal data, tracking individuals over time to assess whether the AI algorithms accurately predict cognitive decline. This data collection is logistically and financially taxing, particularly for startups. Third, once validation is completed, there may be new developments in AI that render the original algorithm outmoded, necessitating further updates and reviews. Thus, it is not unusual to find small companies that have made initial breakthroughs in AI for dementia detection but remain stuck in an endless cycle of development and partial validation, unable to proceed to full-scale implementation.

12.4.2 Robotic Caregivers

Beyond companionship robots like PARO, more advanced robotic caregivers aim to assist with everyday tasks such as feeding, bathing, or facilitating physical therapy. Robots like NAO are used in some clinical settings to guide older adults through simple exercises or cognitive games. The promise is twofold: first, to reduce caregiver burden by automating repetitive tasks, and second, to provide consistent support in settings where human caregivers might be in short supply.

Yet, these advanced robotic solutions face cost and ethical barriers. Development costs are high, often requiring specialized hardware, AI software integration, and extensive engineering to ensure safety. Ethical committees, rightly cautious about possible harm, impose additional safeguards. For instance, a robot that physically interacts with an older adult could inadvertently cause injury if not designed or calibrated properly. Each of these considerations adds to the cost and time required for development, testing, and approval.

Moreover, limited adoption perpetuates a vicious cycle. Because robotic caregivers are expensive, few organizations implement them, which in turn restricts the volume of post-market data needed to refine and validate their real-world performance. As with AI-powered cognitive tools, progress stalls, and many potentially transformative robotic platforms languish in research labs or pilot programs.

The "Island of Misfit Solutions" is a sobering reminder that scientific progress alone does not guarantee broad societal benefit. Many AgeTech concepts that show promise in controlled or early-stage environments fail to clear the multiple hurdles needed for large-scale adoption. In the next section, we explore proposed reforms aimed at dismantling the barriers that produce this phenomenon, offering pathways to accelerate technology deployment without compromising safety, efficacy, or ethical integrity.

12.5 Proposed Reforms for Accelerating Innovation

Addressing the barriers that stymie AgeTech innovation requires systemic changes across regulatory, research, and collaborative domains. The goal is to strike a balance that maintains rigorous standards of safety and efficacy while allowing for the rapid iteration and deployment inherent to emerging technologies. Below are five proposed reforms that, taken together, could significantly streamline the journey from concept to market for AgeTech solutions.

12.5.1 Adaptive Regulatory Frameworks

Regulatory bodies worldwide have begun to experiment with adaptive oversight models that cater to fast-evolving technologies. In the United States, for example, the FDA's Digital Health Innovation Action Plan allows for provisional clearances and iterative review of AI-based healthcare products. This approach acknowledges that AI tools can improve rapidly, and it offers a structured mechanism for updating algorithms without re-entering the full regulatory cycle each time.

Such frameworks could be further expanded to encompass AgeTech devices that interface with older adults in non-traditional ways, such as robotic caregivers or immersive virtual reality systems for cognitive stimulation. By establishing categories for "low-risk" or "moderate-risk" interventions, regulators could fast-track solutions that demonstrate adequate safety measures. Where potential harm is minimal, real-world post-market data could suffice for final certification, streamlining the process while preserving accountability.

12.5.2 Alternative Research Methodologies

Traditional randomized controlled trials are the gold standard for many healthcare interventions, but they may be ill-suited for dynamic, rapidly iterating technologies. As an alternative, real-world evidence (RWE) methodologies can expedite the validation process by collecting data from actual use cases in naturalistic settings [2]. Wearable sensors and electronic health records can yield continuous data, reducing the time and cost of gathering large cohorts. Synthetic datasets, which simulate certain conditions based on existing clinical data, can be used to approximate real-world scenarios without extensive participant recruitment.

Implementing such methodologies, however, requires updated guidelines and widespread acceptance among researchers and regulators. Collaboration between academia, industry, and government bodies could lead to the standardization of RWE metrics for AgeTech. While RCTs remain vital for interventions that could pose significant risks, adopting broader evidence-collection strategies can accelerate innovation in areas like cognitive assessment tools, telehealth, and robotic companions.

12.5.3 Collaborative Models

Public–private partnerships and collaborative research hubs can pool the diverse range of skills needed to bring AgeTech to maturity. Canada's AGE-WELL network stands as a prime example, uniting researchers, startups, established companies, and healthcare professionals under one umbrella. Such consortia enable efficient sharing of resources, expertise, and participant recruitment networks. Additionally, these hubs often facilitate streamlined ethical reviews by creating centralized Institutional Review Boards (IRBs) or specialized ethical committees acquainted with AgeTech-specific challenges.

Collaboration also accelerates the adoption of innovations by fostering relationships between technology developers and end users—healthcare providers, long-term care facilities, and older adults themselves. This user-centric design process ensures that new tools are grounded in real-world needs and that any ethical or usability concerns are identified and addressed early. By distributing costs and risks among multiple stakeholders, collaborative models can mitigate financial barriers that so often hamper small startups.

12.5.4 Streamlined Ethical Oversight

Ethical reviews are essential for protecting vulnerable populations, but they can be more efficiently managed. Centralized or expedited review pathways for low-risk AgeTech studies could significantly reduce delays. For instance, the World Health Organization (WHO) has put forth ethical guidelines specific to aging research, offering a framework for addressing issues like informed consent in populations with cognitive impairments. By aligning IRB processes with these specialized guidelines and creating standardized checklists for common AgeTech scenarios, ethical reviews could be completed more swiftly without compromising participant welfare.

Moreover, policy reforms could promote broader acceptance of remote or tele-ethics reviews, allowing multi-site studies to proceed without logistical delays associated with in-person committee meetings. This is especially relevant as telehealth and other remote interventions gain ground. Streamlining ethical oversight does not mean sidestepping critical protections; rather, it involves optimizing review processes to ensure timely, consistent, and thorough evaluations.

12.5.5 Flexible Evidence Standards

Finally, flexibility in the types of acceptable evidence can bridge the gap between rigorous validation and practical deployment. Regulators like the FDA have begun adopting risk-based approaches, particularly for Software as a Medical Device (SaMD), where lower-risk tools can receive market authorization based on more limited yet compelling evidence. Qualitative data, user testimonials, and observational studies could supplement more traditional quantitative metrics for tools aimed at improving general well-being or mild cognitive support.

Such flexible standards would, for instance, validate a robotic companion's utility if evidence shows it consistently reduces stress and loneliness among older adults, even if large-scale RCT data remain limited. Of course, higher-risk interventions—such as autonomous robots designed for physical transfers—would still require more robust clinical data. Tailoring the type and volume of evidence to the device's risk profile creates a more proportionate and agile system.

Collectively, these five reforms offer a blueprint for dismantling many of the hurdles outlined in earlier sections. By enabling quicker regulatory pathways, diversifying research methodologies, encouraging collaboration, refining ethical oversight, and expanding evidence criteria, we can set the stage for a new era of AgeTech development. Importantly, these reforms must be undertaken with great care to preserve ethical integrity and public trust, which is the focus of the next section.

12.6 Balancing Innovation with Ethical Integrity

While speeding up the pathway to market for AgeTech solutions is vital, it must not come at the expense of ethical considerations. Older adults, particularly those with cognitive impairments or complex health conditions, can be especially vulnerable to exploitation, data misuse, or inadequate care. Consequently, any reforms that relax or modify existing regulations should do so in a manner that safeguards this population's well-being and autonomy. Below, we explore two critical dimensions of ethics in AgeTech: privacy and data security, and human-centered design.

12.6.1 Privacy and Data Security

Many AgeTech solutions rely on continuous monitoring, capturing data such as movement patterns, heart rate, or even emotional states. IoT-enabled devices in smart homes, for instance, collect and transmit real-time information to cloud-based servers for analysis. Telehealth platforms store sensitive medical records that might include diagnosis and treatment data. While such capabilities facilitate personalized and preventive care, they also heighten the risk of data breaches or misuse [4].

Ensuring robust cybersecurity measures is thus a prerequisite for ethical AgeTech deployment. Encryption, secure data transmission protocols, and compliance with data protection regulations like GDPR are essential. However, meeting these standards can be financially and technically challenging for startups, which may be tempted to adopt shortcuts to reduce costs or speed up development. Here, collaborative models and public–private partnerships can offer guidance, shared resources, and cost-effective security solutions that meet regulatory benchmarks. Moreover, these solutions should incorporate user-friendly consent mechanisms, enabling older adults or their caregivers to understand and control how their data are used.

12.6.2 Human-Centered Design

Amid the excitement surrounding robotics and AI, there is a risk of losing sight of the human element. Care is not solely about efficiency or automation; it is also about empathy, companionship, and respecting individual dignity. Over-reliance on technology may inadvertently depersonalize care, especially if robots or AI tools start replacing, rather than augmenting, human caregivers.

A balanced approach, often called a "hybrid model", ensures that robots handle routine, physically demanding, or high-risk tasks while human professionals focus on emotional support and relationship building. This division can improve quality of care, reduce caregiver burnout, and still uphold the interpersonal aspects of eldercare. Human-centered design principles thus call for iterative testing with real users, ensuring that AgeTech is intuitive, dignified, and genuinely responsive to user needs rather than driven solely by engineering constraints.

Ultimately, ethical integrity in AgeTech is not just about avoiding harm—it is about actively promoting well-being, trust, and autonomy among older adults. The next section broadens the conversation by examining how different global contexts, Canada, Japan, the United States, and Europe, approach AgeTech regulation and adoption, illustrating both convergences and divergences that can inform best practices.

12.7 Global Perspectives

AgeTech is a global phenomenon, shaped by diverse healthcare systems, cultural norms, and policy environments. While certain challenges are universal, such as finding a balance between rapid innovation and rigorous validation, different regions have adopted varied strategies to address them. By examining initiatives in Canada, Japan, the United

States, and Europe, we can glean valuable lessons about which approaches might be most successful in catalyzing AgeTech development and ensuring widespread adoption.

12.7.1 Canada

In Canada, the AGE-WELL network stands at the forefront of AgeTech innovation. This nationwide initiative unites universities, healthcare facilities, private sector partners, and community organizations under a single umbrella. Through grants, mentorship programs, and co-development platforms, AGE-WELL seeks to accelerate the creation of technologies that address the urgent needs of older adults. The network also emphasizes interdisciplinary research, bringing together engineers, gerontologists, occupational therapists, and business experts to ensure holistic solutions.

Regulatory-wise, Health Canada has been exploring adaptive pathways similar to those of the FDA. Yet, like many other regions, Canada grapples with the challenge of inconsistent regulations across provinces, particularly regarding telehealth. Still, the collaborative ethos fostered by AGE-WELL and other national research programs demonstrates a successful template for knitting together diverse stakeholders. This collaborative environment not only encourages innovation but also provides robust support for startups facing resource limitations.

12.7.2 Japan

Japan is often cited as a leading indicator of global demographic trends, given its rapidly aging population and shrinking workforce. In response to these demographic pressures, the Japanese government has championed robotics and AI as core components of eldercare solutions. Innovations like Robear—a robot designed to carry or lift older adults—exemplify the country's tech-forward approach. Japan's Ministry of Economy, Trade and Industry (METI) and the Ministry of Health, Labour and Welfare often collaborate on funding initiatives, offering subsidies or tax incentives for companies that develop or deploy caregiving robots.

This proactive government role mitigates some of the resource and regulatory constraints faced by startups in other countries. That said, public acceptance of robots in caregiving roles varies, and concerns about depersonalization persist. Japan's emphasis on robotics also sometimes overshadows development in other areas of AgeTech, such as telehealth or AI-based cognitive assessment tools. Nonetheless, its model illustrates how government-led investment and clear policy direction can galvanize an entire sector, positioning robotics as a viable supplement to human caregiving.

12.7.3 United States

The U.S. Food and Drug Administration (FDA) has increasingly adopted a progressive stance on digital health innovation, as evidenced by initiatives like the Digital Health Innovation Action Plan. By offering pathways for iterative AI development and modular approval processes, the FDA seeks to keep pace with technology's rapid evolution. Federal agencies also fund significant research through grants from entities like the National Institutes of Health (NIH) and the National Science Foundation (NSF), supporting everything from early-stage conceptual research to large-scale clinical trials.

Nevertheless, the United States faces its own set of barriers, particularly around healthcare reimbursement models. Medicare and Medicaid reimbursement policies can be slow to adapt, making it difficult for telehealth platforms or innovative AgeTech devices to secure coverage. Additionally, the country's fragmented insurance landscape complicates the adoption of new technologies, as each insurer may have different standards for coverage. Still, the U.S. market is large and diverse, offering ample opportunities for AgeTech startups—if they can navigate the complexities of reimbursement and regulatory compliance.

12.7.4 Europe

Europe provides an equally diverse landscape, spanning multiple countries with varying degrees of regulatory integration under the European Union (EU). Notably, the EU's General Data Protection Regulation (GDPR) has become a global benchmark for data privacy, influencing the design and deployment of AgeTech solutions worldwide. While GDPR ensures robust protections for sensitive data, it can also pose compliance challenges for startups that must invest in secure data storage and processing infrastructure.

On the innovation side, the European Commission has funded numerous projects focused on Active and Assisted Living (AAL). These programs support cross-border collaborations among member states, encouraging knowledge exchange and co-creation of AgeTech solutions. One challenge, however, is that healthcare remains primarily a national competence within the EU, so each country's regulatory environment and healthcare financing structures can differ significantly. This patchwork system can hinder the scaling of AgeTech solutions across Europe, even if they are proven successful in one member state.

In sum, these global examples highlight a spectrum of approaches to AgeTech—from Japan's robust governmental backing of robotics to Canada's collaborative research networks and the EU's stringent data privacy regime. Each region's experience underscores the importance of tailoring regulatory frameworks, funding mechanisms, and cultural acceptance to the specific context. By drawing insights from these diverse models,

AgeTech stakeholders can forge more integrated and flexible strategies, ideally converging on a global best practices framework that accelerates adoption while safeguarding ethical standards.

12.8 Conclusion

AgeTech stands at a crucial nexus, where rapidly advancing technologies hold the promise of transforming eldercare against the backdrop of a rapidly growing global population of older adults. Robotic companions can combat loneliness and offer emotional support, while AI-driven tools may provide earlier detection of cognitive decline. Smart home systems promise increased autonomy, and telehealth platforms aim to bridge gaps in healthcare access. Yet, for all this potential, traditional demands for rigorous scientific validation, coupled with regulatory complexities and ethical considerations, frequently slow the translation of these innovations from concept to widespread implementation.

This essay has illustrated how each aspect of the AgeTech ecosystem, be it robotic caregiving, AI-based cognitive tools, or telehealth services, contends with a set of obstacles. Extended clinical trials and validation processes, although vital for safety, can render technologies obsolete before they reach the market. Smaller innovators with limited resources often find themselves at a disadvantage, unable to invest the capital or time needed for comprehensive studies. On the regulatory front, frameworks initially designed for conventional medical devices can be ill-suited for rapidly evolving AI or robotic systems, thus creating uncertainty and further deterring investment. Ethical and legal dimensions, including data privacy and the necessity of informed consent, add additional layers of complexity, particularly when older adults with cognitive impairments are involved.

Yet, these challenges, while formidable, are not insurmountable. Adaptive regulatory frameworks that allow for provisional approvals can help keep pace with swift technological evolution. Alternative research methodologies, such as real-world evidence or synthetic data simulations, may offer more efficient routes for generating credible results. Collaborative models, exemplified by networks like AGE-WELL, demonstrate the power of uniting diverse stakeholders to pool resources and expertise. Streamlined ethical oversight, relying on specialized guidelines and centralized IRBs, has the potential to safeguard older adults while reducing approval times. Finally, adopting flexible evidence standards, especially for lower-risk interventions, can maintain essential rigor without stifling creativity.

Maintaining ethical integrity, however, remains paramount. Older adults constitute a heterogeneous population with varied health statuses, living conditions, and personal preferences. Designing technologies that genuinely enhance well-being must involve careful consideration of privacy, data security, and respect for personal autonomy. Human-centered design principles insist that technology should complement, rather than supplant, the warmth and empathy found in human caregiving interactions.

Looking across different global contexts reveals both uniformities and unique strategies. Japan's government-led initiatives in robotics, Europe's stringent data privacy regulations, the FDA's evolving digital health policies in the United States, and Canada's collaborative research networks all point to the multifaceted ways nations are adapting to an aging society. These approaches underscore the value of international collaboration, knowledge exchange, and policy harmonization in forging a path that harnesses AgeTech's potential to the fullest.

In closing, the future of AgeTech depends on our collective ability to reconcile the need for innovation with the imperatives of safety, efficacy, and respect for human dignity. Rapid demographic changes demand urgent solutions, yet rushing technologies to market without sufficient oversight risks harming those they aim to serve. Achieving the correct balance involves re-envisioning peer review, regulatory, and ethical frameworks to align more closely with the unique characteristics of AgeTech. By embracing nuanced reforms, such as adaptive approval pathways, real-world evidence methodologies, collaborative research networks, efficient ethical review processes, and flexible yet robust evidence standards, we can pave the way for AgeTech solutions that truly meet the needs of a growing aging population. It is through this delicate equilibrium that society can ensure that the undeniable promise of AgeTech is realized, ushering in a future where older adults can lead healthier, more independent, and more fulfilling lives.

References

1. Fraser, K. C., Meltzer, J. A., Rudzicz, F., & Garrard, P. (2016). Linguistic Features Identify Alzheimer's Disease in Narrative Speech. *Journal of Alzheimer's Disease, 49*(2), 407-422. https://doi.org/10.3233/jad-150520
2. Sherman, R. E., Anderson, S. A., Dal Pan, G. J., Gray, G. W., Gross, T., Hunter, N. L., Robb, M. A. et al. (2016). Real-world evidence—what is it and what can it tell us. *New England Journal of Medicine, 375*(23), 2293–2297.
3. United Nations. (2024). World Population Prospects 2024. Retrieved from https://www.un.org/development/desa/pd/world-population-prospects-2024
4. Ziegeldorf, J. H., Morchon, O. G., & Wehrle, K. (2014). Privacy in the Internet of Things: threats and challenges. *Security and Communication Networks, 7*(12), 2728–2742. https://doi.org/10.1002/sec.795

References

13 Conducting Research in AgeTech: Paradigms, Approaches, and Challenges

13.1 Introduction

The global aging population is one of the most pivotal demographic developments of the twenty-first century, with far-reaching consequences that cut across economic, social, and healthcare domains. Projections indicate that by 2050, the number of individuals aged 60 and older will surpass 2.1 billion, representing over 20% of the global population [3]. This demographic shift is multifaceted, offering an array of opportunities and challenges to policymakers, healthcare providers, researchers, and industries that cater to older adults. On one hand, an older population contributes accumulated wisdom, experience, and economic potential, as many seniors continue to work, innovate, and influence social structures. On the other hand, an aging society also places significant demands on healthcare systems, social welfare programs, and the broader economy. Issues such as chronic disease management, social isolation, and caregiving complexities require innovative and evidence-based approaches to ensure that older adults maintain a high quality of life.

In this context, AgeTech has emerged as a transformative domain that encompasses various innovations, including wearable sensors, robotic caregiving devices, telemedicine platforms, and smart home technologies, specifically aimed at improving the lives of older adults. Unlike general-purpose technologies, AgeTech solutions are designed to meet unique age-related requirements such as reduced mobility, cognitive changes, and the increased prevalence of comorbidities. These technologies, if properly developed and integrated, can promote aging in place, enhance social participation, and alleviate burdens on formal and informal caregivers.

However, realizing the potential of AgeTech is contingent on the robustness and relevance of research methodologies employed to develop and evaluate these innovations. While traditional research paradigms, such as quantitative clinical trials, remain essential,

A. Mihailidis, *The Age of AgeTech*, Synthesis Lectures on Technology and Health,
https://doi.org/10.1007/978-3-032-04886-8_13

AgeTech also demands approaches that account for older adults' lived experiences, varying abilities, and the societal contexts in which technologies are used. Researchers must navigate a complex landscape that involves not only scientific and technical rigor but also ethical principles, accessibility standards, and user-centered design processes.

Moreover, the interdisciplinary nature of AgeTech research necessitates collaboration among technologists, healthcare providers, social scientists, designers, and end users. Such collaborations help ensure that new interventions are not merely technologically sophisticated but also meaningful, usable, and equitable. The importance of ethical and regulatory considerations cannot be overstated: data privacy, especially concerning sensitive health information, emerges as a central issue. Furthermore, research methods must adapt to rapid technological change, as tools tested under certain conditions can become obsolete before they are widely disseminated.

This essay explores the paradigms, practices, and principles that guide AgeTech research, laying out a roadmap for how researchers, practitioners, and policymakers can advance the field responsibly and effectively. Beginning with an overview of quantitative, qualitative, and mixed-methods research paradigms, the discussion underscores the importance of user-centered design in ensuring that solutions align with older adults' actual needs. We then delve into key challenges, such as recruitment barriers, ethical dilemmas, and fast-paced technological changes. The essay also illustrates how interdisciplinary collaboration enriches AgeTech research, incorporating essential insights from engineering, healthcare, and social sciences. Next, we examine real-world evidence and its significance in supplementing or, in some cases, supplanting traditional clinical trials. We then take a global view, highlighting initiatives in Canada, Japan, the United States, and Europe to illustrate varying research and policy frameworks. Finally, this essay points toward the future of AgeTech research, emphasizing the potential of AI-driven tools, cross-cultural studies, and hybrid research methodologies to transform how we develop technologies for an aging world.

By presenting a comprehensive, in-depth look at AgeTech research paradigms and practices, this essay seeks to encourage the adoption of evidence-based, ethically grounded, and user-focused approaches. As societies around the world continue to grapple with the realities of aging populations, rigorous and inclusive AgeTech research will remain a linchpin for ensuring that emerging technologies genuinely benefit older adults, caregivers, and communities. The decisions made today, from forming cross-sector collaborations to embracing participatory methods, will shape tomorrow's landscape of tools, policies, and social attitudes toward aging. Hence, the following sections provide a detailed exploration of each aspect of AgeTech research, offering both a roadmap and a rationale for how best to navigate this rapidly evolving field.

13.2 Research Paradigms in AgeTech

13.2.1 Quantitative Research Paradigms

Quantitative research paradigms play a critical role in establishing both the efficacy and safety of AgeTech solutions. These paradigms are built on structured data collection, hypothesis testing, and statistical analysis. By systematically measuring and analyzing variables, researchers can draw objective conclusions about a technology's performance, paving the way for evidence-based decision-making among clinicians, caregivers, policymakers, and end users. Several quantitative research designs are particularly relevant:

- **Randomized Controlled Trials (RCTs).** RCTs are often described as the gold standard for evaluating interventions because they minimize bias through random assignment and the inclusion of control groups. In AgeTech, RCTs might be used to test a new wearable device designed for detecting falls in older adults. Researchers randomly assign participants to either use the new device or receive usual care (or an alternative technology) and then compare outcomes such as fall rates and user satisfaction. The advantage of RCTs lies in their ability to draw clear causal inferences: if a difference is observed between the intervention and control groups, researchers can be more confident that the AgeTech innovation caused the difference. A notable example is the validation of wearable fall detection systems, where RCTs have demonstrated enhanced accuracy and reliability [2]. Despite their strengths, RCTs can be resource-intensive, requiring substantial funding and extensive participant recruitment, which can be challenging when working with older populations who may have mobility limitations or comorbid conditions.
- **Cross-Sectional and Longitudinal Studies.** These types of studies provide complementary insights into the performance and impact of AgeTech solutions over time. In cross-sectional designs, data are collected at one point to assess the prevalence of certain behaviors, usage patterns, or attitudes toward technology. This snapshot can highlight immediate concerns such as usability bottlenecks or adoption rates across different demographic segments. Longitudinal studies, in contrast, track participants over months or even years, offering insights into how AgeTech influences, or is influenced by, changes in health status, technological familiarity, and social dynamics. Such studies can help identify long-term benefits or drawbacks of particular interventions. For example, a longitudinal study could track older adults' use of a telehealth platform to monitor chronic conditions, examining how usage correlates with health outcomes, hospital readmissions, or quality-of-life measures over several years.

- **Experimental and Quasi-Experimental Designs**. Beyond RCTs, other quantitative designs like quasi-experimental approaches (e.g., time-series analyses, matched comparisons) can still yield valuable insights without the strict requirements of randomization. These designs are particularly useful in community or clinical settings where random assignment is difficult or ethically complex, such as when an AgeTech tool is introduced across an entire residential care facility rather than individual residents.

By applying these quantitative research paradigms, AgeTech researchers can move beyond anecdotal evidence to establish a scientific basis for technology adoption. Quantitative findings can guide policymakers in allocating resources and inform further refinements of the technology design, ultimately leading to solutions that more effectively address the needs of older adults.

13.2.2 Qualitative Research Paradigms

While quantitative research focuses on measurement and causal relationships, qualitative paradigms provide depth of understanding about user experiences, motivations, and perceptions. In AgeTech, where individual needs can vary greatly due to differences in physical, cognitive, and emotional conditions, qualitative insights are indispensable for capturing the nuances of technology interaction and adoption. Two prominent qualitative approaches include.

- **Interviews and Focus Groups**. In-depth interviews and focus group discussions allow researchers to explore older adults' perspectives on technology use. For instance, an interview might investigate how an older adult perceives a newly introduced wearable health monitor—probing what they find helpful, what challenges they face, and how they believe it affects their sense of independence. Focus groups, on the other hand, encourage interactive dialogue among several participants, enabling the emergence of collective themes such as shared frustrations or positive experiences. These methods have been utilized by organizations like AGE-WELL in Canada, which has conducted focus groups to inform the design and refinement of telehealth platforms. By engaging participants in open-ended discussions, researchers can capture subtle emotional responses, concerns about privacy, or recommendations for improving usability.
- **Ethnographic Studies**. Ethnography involves observing participants in their natural environments, be it their home, community centers, or healthcare facilities, over an extended period. This approach is particularly effective for identifying real-world opportunities and constraints that might not be evident in controlled settings. In the context of AgeTech, ethnographic fieldwork can reveal how older adults navigate technology during daily routines, whether they rely on caregivers or family members for

assistance, and what barriers to adoption might arise due to architectural constraints or sociocultural factors. Such insights often highlight factors like stigma associated with assistive devices, the emotional comfort or discomfort derived from robotic caregivers, or resistance to adopting new routines that disrupt familiar patterns.

Qualitative research not only uncovers the "how" and "why" behind user behavior but also lays the foundation for design improvements. For instance, if interviews repeatedly reveal concerns about data privacy, designers can prioritize the development of more transparent data-handling features. If ethnographic observations show that older adults rarely wear wristband devices in certain climates, alternative form factors like clip-on sensors or footwear-based sensors might be considered. In sum, qualitative methods ensure that AgeTech solutions resonate with older adults' lived experiences rather than just meeting technical performance metrics.

13.2.3 Mixed-Methods Approaches

Mixed-methods research merges the strengths of both quantitative and qualitative paradigms, offering a holistic understanding of how AgeTech innovations perform across multiple dimensions. This approach is especially pertinent in evaluating highly complex interventions, where technology efficacy must be assessed in tandem with user acceptance, ethical considerations, and contextual factors. Creswell and Clark [1] argue that mixed-methods designs integrate statistical analyses with rich qualitative data, enabling the triangulation of findings and producing more robust conclusions [1].

A typical mixed-methods study in AgeTech might involve an initial quantitative phase, such as an RCT to evaluate the accuracy of a fall detection device, followed by a qualitative phase that explores participants' subjective experiences using that device. Researchers may distribute validated instruments like the System Usability Scale (SUS) or the Technology Acceptance Model (TAM) questionnaires, yielding quantitative scores that can be statistically analyzed. Complementary focus groups or interviews can then explore why certain participants scored the device as easy to use while others struggled, revealing the underlying reasons for acceptance or resistance.

Another configuration of mixed-methods applies an exploratory sequential approach, beginning with qualitative data collection, such as ethnographic observations in a senior care facility, to identify potential challenges or feature requirements. These insights then inform the development of a quantitative instrument, like a survey measuring older adults' confidence in using technology or their perceived improvements in daily functioning. Ultimately, the synergy of quantitative rigor and qualitative depth ensures that AgeTech research speaks to both objective performance metrics and subjective human experience.

13.3 User-Centered Design in AgeTech Research

13.3.1 Principles of User-Centered Design

User-centered design (UCD) is a cornerstone in the development of AgeTech solutions, emphasizing that older adults—and in many cases their caregivers—must be actively involved in each stage of the design process. UCD aims to align technological innovation with the real-world needs, abilities, and preferences of end users, rather than expecting users to adapt to technology that may not be accessible or intuitive.

- **Involvement of End Users**. UCD frameworks encourage continuous collaboration with older adults. Their direct participation in ideation sessions, prototyping, and testing allows designers to capture critical insights early. Instead of relying solely on assumptions about what older adults might need, developers gain firsthand feedback on design elements such as screen layout, font sizes, and color contrast for readability. This reduces the risk of launching a product that might be functionally robust but fails to resonate with its target audience.
- **Iterative Design**. Iteration involves creating prototypes or beta versions of an AgeTech tool, testing them with users, analyzing feedback, and refining the design in multiple cycles. This process reduces the likelihood of major usability issues going undetected until late in development, when they are more expensive to fix. Iterative design also allows for incremental improvements, enabling designers to explore multiple design alternatives and select the most user-friendly option.
- **Accessibility and Simplicity**. Accessible design principles go beyond providing larger font sizes. They may include voice-activated commands, high-contrast color palettes, or tactile feedback. By incorporating these accessibility features, AgeTech solutions can account for common age-related changes such as reduced vision, hearing loss, and diminished fine motor control. Simplicity likewise ensures that older adults are not overwhelmed by complex menus or excessive options, promoting ease of navigation and user satisfaction.

Ultimately, UCD is rooted in empathy. It compels designers and researchers to walk in the shoes of older adults, acknowledging that technology can be a double-edged sword—empowering in some cases, daunting in others. By foregrounding the perspectives of end users, designers are more likely to produce tools that genuinely enhance quality of life rather than merely adding another layer of complexity.

13.3.2 Participatory Research

Participatory research approaches extend beyond traditional UCD by positioning older adults as co-researchers rather than passive subjects. This approach is grounded in the belief that individuals who are directly affected by a problem or need, such as mobility constraints, risk of social isolation, or chronic health issues, hold valuable expertise that can guide research and development. In practice, participatory research can take many forms:

- **Co-Design Workshops**. These workshops invite older adults, caregivers, clinicians, and researchers to engage collectively in brainstorming and prototyping sessions. Using paper sketches, storyboards, or mock-ups, participants articulate their needs, evaluate proposed solutions, and suggest improvements based on their lived experiences. This collaborative environment fosters mutual learning: while designers gain insights into daily challenges faced by older adults, participants learn about potential technological capabilities and constraints.
- **Action Research Projects**. In action research, researchers and participants collaborate to diagnose issues, design interventions, and evaluate outcomes in real-world settings. This iterative cycle of "planning, acting, observing, and reflecting" helps ensure that solutions remain relevant over time. For example, Canada's AGE-WELL network has championed participatory research models by working closely with seniors to co-create smart home systems, wearable sensors, and cognitive health applications. The outcome is not merely a single product but an ongoing exchange of ideas, leading to more adaptable, context-specific solutions.
- **Empowerment and Ownership**. A key benefit of participatory methods is that older adults gain a sense of empowerment and ownership. Rather than feeling like subjects in a controlled study, participants become active contributors shaping the research agenda, design choices, and evaluation metrics. This can also boost engagement and reduce attrition rates, as individuals are more likely to remain committed to projects that directly incorporate their voices and perspectives.

Participatory research thus enriches AgeTech design by grounding innovations in the practical realities of aging, ensuring that final products address actual needs rather than perceived ones. Simultaneously, it builds community engagement and potentially fosters wider acceptance of technologies once they are released.

13.3.3 Usability Testing

Usability testing is a crucial checkpoint in the development of AgeTech solutions, aimed at evaluating how easily and effectively older adults can interact with a technology in practice. Even if a device or software is theoretically designed for usability, real-world testing often reveals unforeseen pain points, accessibility issues, or user confusion.

- **Laboratory versus Field Usability Testing**. In laboratory testing, participants are asked to complete specific tasks under controlled conditions while researchers observe or record their interactions. This environment allows for detailed data collection—such as the exact time it takes to complete a task or the number of errors. However, controlled settings may not fully capture the natural contexts in which older adults use technology. Field testing, conducted in real environments like homes or community centers, offers more ecological validity, although it can introduce additional variables that complicate data interpretation.
- **Evaluation Metrics and Tools**. Tools such as the System Usability Scale (SUS), task completion rates, and error logs quantify user performance. Meanwhile, qualitative feedback, collected through post-task interviews, adds context by clarifying why certain elements were intuitive or confusing. For instance, an older adult might struggle with a touchscreen interface not because of poor design but because of arthritis-related difficulty applying enough pressure to the screen.
- **Case Example: GrandPad Tablet (U.S.)**. The GrandPad was specifically designed to cater to seniors by featuring larger icons, simplified navigation, and one-touch connectivity to family members. Usability testing with older adults of varying technological proficiencies was instrumental in refining these features. Test sessions revealed potential pitfalls, such as the need for better glare reduction and simplified onboarding. These insights led to iterative improvements that made the GrandPad more approachable and user-friendly for its target audience.

Through rigorous usability testing, AgeTech developers can pinpoint design flaws early, refine the user interface, and optimize accessibility features. In turn, the iterative improvements that follow help ensure that the final product is genuinely usable and beneficial, thereby increasing the likelihood of sustained adoption.

13.4 Challenges in AgeTech Research

13.4.1 Recruitment and Retention

One of the most persistent challenges in AgeTech research is recruiting and retaining older adults as study participants. Despite growing awareness about the potential benefits of AgeTech, many older adults remain skeptical or anxious about engaging with new technologies, particularly if they have limited prior experience. Several strategies can help overcome these hurdles:

- **Incentives and Engagement**. Offering tangible incentives, such as financial compensation or free access to AgeTech devices, can encourage participation. Yet non-monetary benefits, like opportunities to socialize, learn new skills, or contribute to research that may improve the lives of their peers, can also be highly motivating. When older adults perceive genuine value in the study, they are more likely to remain committed and provide thoughtful feedback.
- **Community Partnerships**. Collaborating with organizations that already serve older adults, such as senior centers, faith-based groups, or advocacy organizations, can streamline recruitment. These partners often have built-in trust and credibility within local communities. Researchers can host informational sessions at community venues, demonstrating the technology and clarifying any risks or benefits. This approach reduces barriers by meeting older adults in familiar, accessible environments and leveraging pre-existing social networks.
- **Flexible Study Designs**. Older adults may face mobility limitations, caregiving responsibilities, or chronic health issues that affect their availability to participate in research activities. Designing studies that accommodate these realities, through flexible scheduling, remote data collection, or transportation assistance, can help retain participants over the long term. Periodic phone calls or digital check-ins can also maintain engagement between formal assessment periods.

Recruitment and retention challenges are not merely logistical; they also have ethical and scientific implications. If the sample of older adults is not diverse in terms of health status, socioeconomic background, or cultural context, the resulting data may lack generalizability. Strategies that proactively address these barriers can yield a more representative and robust evidence base for AgeTech innovations.

13.4.2 Ethical Considerations

Ethical issues loom large in AgeTech research, given that interventions often involve sensitive health data, direct interactions with vulnerable populations, and technology that can impact individuals' autonomy or privacy. Key ethical considerations include.

- **Informed Consent**. Ensuring that older adults fully understand the nature and purpose of a study, along with any associated risks, is paramount. This is especially critical for participants with mild cognitive impairment or sensory limitations that may impede comprehension. Researchers might employ simplified consent forms, verbal explanations, or visual aids to clarify the study's scope and requirements. Ongoing consent is equally important, allowing participants to withdraw at any stage without penalty.
- **Privacy and Security**. Many AgeTech solutions, ranging from wearable health monitors to home-based sensors, collect continuous data on activity levels, vital signs,

or even location. This raises significant questions about data protection. Researchers and developers must implement secure protocols for data storage, transmission, and analysis, in line with established frameworks like the EU's General Data Protection Regulation (GDPR). For instance, [4] emphasize that IoT-based healthcare devices must incorporate encryption, secure authentication, and minimal data retention to protect users' sensitive information [4].

- **Autonomy versus Safety**. AgeTech frequently aims to mitigate safety risks, such as falls or medication non-adherence. However, overly intrusive monitoring tools can diminish users' sense of autonomy or privacy. A device that constantly tracks an older adult's location might be reassuring for family members but could be perceived as invasive by the user. Ethical research design must strike a balance between leveraging technology for improved care and safeguarding individuals' dignity and independence.

Addressing these ethical considerations is not just a regulatory requirement but a moral imperative that underscores respect for older adults as autonomous agents. Researchers who integrate ethical frameworks from the outset are more likely to develop AgeTech solutions that are both efficacious and aligned with the values of end users.

13.4.3 Technological Adaptation

The rapid pace of technological change presents another complex challenge in AgeTech research. Studies that span multiple years may find that the hardware or software being tested becomes outdated or discontinued before the trial concludes. This can disrupt data collection, undermine the relevance of results, and frustrate both researchers and participants.

- **Iterative and Adaptive Protocols**. One solution involves designing protocols that are flexible enough to incorporate updates or improvements in the technology. Instead of locking down the technology specification for the entire duration of a study, researchers can allow for incremental upgrades, provided they do not compromise the study's scientific integrity. Pilot phases can also help identify potential vulnerabilities in technology design before large-scale implementation.
- **Technology-Agnostic Outcome Measures**. By focusing on outcome measures that are less dependent on a specific hardware or software platform, researchers can future-proof their studies. For instance, instead of evaluating a particular brand of robotic caregiver, researchers might measure variables like user satisfaction with personal assistance, improvements in mobility, or reductions in caregiver burden. This enables the study's findings to remain relevant even if the specific robotic device changes.
- **Cross-Disciplinary Collaboration**. Partnering with technology developers, engineers, and industry stakeholders can help research teams stay abreast of upcoming changes

or new product releases. Early communication about product life cycles, expected updates, and potential obsolescence can guide study design to be more adaptive and less vulnerable to shifts in the technology landscape.

Addressing technological adaptation is crucial if AgeTech research is to remain timely and useful. A dynamic, forward-thinking approach ensures that studies continue to inform best practices, guiding the development of future generations of AgeTech tools and ultimately benefiting older adults through more refined, evidence-based solutions.

13.5 Interdisciplinary Collaboration in AgeTech Research

AgeTech research resides at the intersection of multiple fields, including engineering, healthcare, social sciences, and policy. Effective collaboration across these disciplines is crucial to create technologies that are not only technologically sound but also socially, ethically, and clinically appropriate. Each discipline contributes unique perspectives, shaping innovations that are more holistic and adaptable to the multifaceted needs of older adults.

13.5.1 Engineering and Design

Engineers and designers bring expertise in product development, human–machine interfaces, and software/hardware integration. In AgeTech, this role might involve designing sensor systems for fall detection, developing algorithms for personalized care, or creating mechanical and electronic frameworks for robotic caregivers. However, engineering and design cannot occur in a vacuum; feedback from older adults, clinicians, and caregivers is essential. Collaborative design processes that include iterative feedback loops are often the key to achieving practical, user-friendly solutions. By working directly with healthcare providers or social scientists, engineers gain a deeper understanding of issues like infection control, cognitive impairment, or housing constraints that could influence a product's feasibility.

13.5.2 Healthcare

Healthcare professionals, ranging from nurses and occupational therapists to geriatricians and physiotherapists, are on the front lines of patient care and have direct insights into patient needs, clinical workflows, and therapeutic outcomes. Their input helps to validate the clinical relevance of new AgeTech solutions. For instance, a robotic caregiver might be designed to support transfers from bed to wheelchair, but without input from physiotherapists, the device may not align with safe patient-handling protocols. Collaboration ensures that technologies address actual clinical pain points and can be smoothly integrated into everyday caregiving practices. Japan's development of robotic caregivers like Robear illustrates how healthcare professionals can guide the design and testing of assistive robots ensuring their functions align with real-world healthcare settings.

13.5.3 Social Sciences

Sociologists, psychologists, anthropologists, and other social scientists delve into the human factors underpinning technology acceptance, social relationships, and cultural attitudes. Their work sheds light on how older adults interpret and navigate new technologies, including issues of stigma, trust, and intergenerational communication. By analyzing social structures, such as family dynamics or community resources, social scientists inform interventions that are culturally sensitive and emotionally resonant. They may evaluate how AgeTech influences the mental well-being of older adults or whether it exacerbates or reduces social isolation. This perspective is critical for ensuring that even the most advanced AgeTech tools align with the social realities of aging.

Interdisciplinary collaboration offers a powerful synergy: engineering and design ensure robust functionality; healthcare professionals validate clinical or care-related efficacy; and social scientists refine the user experience and broader societal impact. When policy experts, economists, and ethicists also join the conversation, the resulting AgeTech solutions are likely to be both effective and sustainable. This synergy underlines the importance of building research consortia and networks that span multiple domains, ensuring that each innovation receives the comprehensive scrutiny it requires before widespread adoption.

13.6 Real-World Evidence in AgeTech Research

13.6.1 Observational Studies

Traditional clinical trials, while vital, may not fully capture the complexities and variances of everyday life, particularly for older adults who have diverse living conditions, health statuses, and social networks. Real-world evidence (RWE) methodologies step into this gap, examining how AgeTech solutions function beyond controlled lab environments. One of the core methods involves observational studies, which monitor participants in their natural settings—be it at home, in assisted living facilities, or in community centers. By doing so, researchers can identify how older adults genuinely interact with technology, whether they follow recommended usage guidelines, and how environmental factors (like home layouts or social support structures) affect adoption. This situational clarity can reveal small but significant details that might otherwise go unnoticed, such as a senior's reluctance to consistently wear a smart wristband because it conflicts with personal fashion preferences or causes skin irritation in warmer climates.

13.6.2 Wearable Data Collection

Wearable devices represent a rapidly evolving frontier in AgeTech, with products ranging from simple step counters to sophisticated multi-sensor systems that track heart rate, sleep patterns, and even stress levels. These devices can generate copious amounts of data, allowing researchers to analyze trends, detect anomalies, and measure the efficacy of interventions in real time. For instance, a wearable ECG monitor by a company such as Biotricity could continuously track cardiac rhythms, sending alerts if arrhythmias are detected. In addition to providing immediate health benefits, these real-time data streams enable researchers to study patterns of behavior, such as daily activity levels or sleep disturbances, and correlate them with clinical outcomes. This can lead to early interventions or adjustments in care plans, ultimately improving health outcomes for older adults.

13.6.3 Implementation Science

Implementation science plays a pivotal role in translating RWE into tangible practice changes. While identifying an effective AgeTech tool is one part of the puzzle, ensuring its seamless integration into existing care systems is another challenge entirely. Implementation science examines how organizational structures, care protocols, workforce training, and patient attitudes either facilitate or hinder new technology adoption. Through methods like process evaluations or cost-effectiveness analyses, researchers identify bottlenecks, be they related to staff hesitancy, insufficient infrastructure, or limited reimbursement models, and propose strategies to overcome them. For instance, if a hospital system wants to adopt a telehealth platform for remote consultations, implementation science can help map out necessary changes in workflow, training needs for nurses and physicians, and strategies to engage older adults who may be unfamiliar with video conferencing tools.

By focusing on real-world conditions, RWE approaches, including observational studies, wearable data collection, and implementation science, help researchers go beyond proof of concept to achieve practical impact. This comprehensive perspective ensures that AgeTech innovations do not merely exist on paper or in isolated pilot studies but are successfully scaled, sustained, and integrated into the care continuum.

13.7 Future Directions in AgeTech Research

13.7.1 AI-Driven Research Tools

Artificial intelligence (AI) offers immense possibilities for enhancing AgeTech research and interventions. Advanced machine learning algorithms can process large, complex datasets, such as continuous sensor data from wearables or electronic health records, and identify patterns indicative of health risks or declines in function. AI-driven analytics can accelerate the research cycle by automating tasks like data cleaning, anomaly detection, and predictive modeling. For example, an AI system could automatically flag early signs of cognitive decline by analyzing speech patterns or usage logs from a digital platform. However, the deployment of AI raises new challenges in transparency, bias, and accountability. Researchers must ensure that AI models are trained on diverse datasets representative of older adults from various backgrounds to avoid perpetuating biases in healthcare outcomes.

13.7.2 Cross-Cultural Studies

As global aging intensifies, AgeTech solutions must be culturally adaptable. Cross-cultural studies can reveal how older adults in different regions interpret and adopt technology, shedding light on variations driven by language, tradition, social norms, and economic constraints. These studies might compare user acceptance of a telehealth platform in rural versus urban settings or across countries with distinct healthcare systems. By exploring these divergences, researchers can refine prototypes that accommodate specific local nuances—such as language translations, user interface modifications, or integration with existing community networks. Cross-cultural research also fosters international collaboration, enabling the transfer of best practices and innovations across borders. When solutions are co-developed with diverse populations, they are more likely to have global relevance and equitable impact.

13.7.3 Hybrid Research Models

The future of AgeTech research likely lies in hybrid research models that blend traditional clinical trials with real-world evidence and participatory methods. Such models can capture both the controlled rigor of randomized trials and the contextual richness of field-based observations. For instance, a hybrid study might start with a pilot RCT in a hospital setting to test a new sensor's efficacy in detecting adverse events, followed by deployment in various community settings for a real-world usage phase. Participants, caregivers, and clinicians would then co-create modifications to address challenges identified in the field. This iterative cycle links experimental evaluation with practical implementation, producing a deeper, more actionable evidence base.

Hybrid models also resonate with the push toward personalized medicine. As big data analytics become more sophisticated, AgeTech solutions can be customized to align with individual health profiles and lifestyles. The combination of robust trials, user feedback, AI-driven insights, and real-world usage data can converge to create adaptive technologies that evolve in tandem with the user. This approach not only accelerates the research process but also ensures that final products are more closely aligned with actual needs, fostering long-term acceptance and improving outcomes for older adults.

13.8 Conclusion

Conducting research in AgeTech is both a challenge and an opportunity—one that demands rigorous methods, inclusive design processes, and a commitment to ethical responsibility. As the global population continues to age, the stakes grow higher for developing innovations that can enhance health, safety, and social engagement for older adults. Throughout this essay, we have examined a spectrum of research paradigms, quantitative, qualitative, and mixed-methods, that provide a comprehensive toolkit for evaluating the effectiveness and acceptance of emerging technologies.

Central to this endeavor is the principle of user-centered design, which ensures that older adults are not peripheral subjects but active participants in shaping technological solutions. Participatory research models further deepen this engagement by involving older adults as co-researchers, thus infusing every stage of the research and design process with lived expertise. However, AgeTech research does not occur in a vacuum: challenges such as participant recruitment, ethical concerns around informed consent and data privacy, and the need to keep pace with rapid technological change all require thoughtful strategies and adaptive protocols.

Interdisciplinary collaboration is another hallmark of successful AgeTech research. By uniting the expertise of engineers, healthcare professionals, social scientists, and end users, researchers can build solutions that are robust, clinically relevant, and socially attuned. Equally important is the shift toward real-world evidence, which bridges the gap between controlled trials and the messy realities of everyday life. Observational studies, wearable data collection, and implementation science collectively offer a more nuanced view of how older adults actually engage with technology, revealing both enablers and barriers to adoption.

Internationally, AgeTech research takes on different nuances shaped by cultural, regulatory, and infrastructural contexts. Canada's AGE-WELL network highlights the power of participatory and user-focused initiatives; Japan's government and private sector collaborations underscore the potential of robotics in an aging society; the United States showcases a regulatory environment increasingly supportive of digital health innovations;

and Europe demonstrates an unwavering commitment to ethical and privacy considerations through frameworks like the GDPR. These global perspectives converge on the shared goal of empowering older adults, underscoring that while paths may differ, the destination remains the same.

Looking to the future, AI-driven analytics, cross-cultural studies, and hybrid research models stand out as key frontiers in AgeTech research. AI promises to rapidly accelerate data processing and predictive modeling, albeit with cautionary notes around bias and transparency. Cross-cultural studies will enrich our collective understanding of how social norms and cultural values affect technology adoption, thereby enabling the creation of more inclusive designs. Meanwhile, hybrid models hold the promise of combining the best of controlled experimentation with the authentic insights of real-world usage, producing technologies that are both rigorously validated and authentically applicable.

In sum, AgeTech research is poised at a critical juncture where scientific rigor meets ethical obligation and technological innovation meets human need. The methodologies and frameworks outlined in this essay aim to guide researchers, designers, and policymakers in navigating this dynamic landscape. By prioritizing user-centered design, interdisciplinary collaboration, and robust, ethically grounded methods, the AgeTech community can harness the immense potential of emerging technologies to foster independence, health, and well-being in older adulthood. The work that unfolds in the coming years will shape how societies worldwide adapt to the realities of an aging population, ensuring that technology becomes a powerful ally rather than a barrier for older adults.

References

1. Creswell, J. W., & Clark, V. L. P. (2017). *Designing and conducting mixed methods research*: Sage publications.
2. Sherman, R. E., Anderson, S. A., Dal Pan, G. J., Gray, G. W., Gross, T., Hunter, N. L., Robb, M. A. et al. (2016). Real-world evidence—what is it and what can it tell us. *New England Journal of Medicine, 375*(23), 2293-2297.
3. United Nations. (2024). World Population Prospects 2024. Retrieved from https://www.un.org/development/desa/pd/world-population-prospects-2024
4. Ziegeldorf, J. H., Morchon, O. G., & Wehrle, K. (2014). Privacy in the Internet of Things: threats and challenges. *Security and Communication Networks, 7*(12), 2728-2742. https://doi.org/10.1002/sec.795

14 Why I Got into the AgeTech Field: A Personal Journey

14.1 Introduction

When I reflect on my career, it is clear that my journey into AgeTech was not straightforward but a confluence of personal experiences, serendipitous encounters, and a deeply ingrained desire to make a difference. From my early research into artificial intelligence (AI) and human–computer interaction to my role as a caregiver for my aging parents and my personal experience of aging with a disability following a spinal cord injury, my path has been shaped by challenges and opportunities. This essay recounts the milestones of my career, starting with a conversation that planted the seed of my passion for AgeTech and exploring how each chapter of my life has reinforced my commitment to this field.

14.2 The Encounter that Changed Everything

In the early days of my career, I had the privilege of meeting an engineer whose wife was living with early-onset dementia. During a candid discussion, he shared the profound opportunities and challenges of caregiving, a role he had taken on while simultaneously managing the emotional weight of watching his wife's abilities decline. One story struck me in particular: his wife often needed help using the toilet, and he had to guide her step by step through the process. He spoke about the embarrassment and discomfort this caused them both, eroding her dignity and increasing his emotional burden.

At one point, he looked at me and asked a simple but profound question: "Wouldn't it be great if computers could do this for us?" That moment changed everything. His words planted a seed in my lifelong pursuit of developing technologies to support older adults and their caregivers. I began to explore how computers and AI could be used to create prompting technologies that would help individuals with dementia maintain their

A. Mihailidis, *The Age of AgeTech*, Synthesis Lectures on Technology and Health,
https://doi.org/10.1007/978-3-032-04886-8_14

independence during self-care activities. This meeting sparked a research interest and a deep sense of responsibility to explore the fundamental and often overlooked struggles caregivers and older adults face.

14.3 Exploring Prompting Technologies and AI

Motivated by this encounter, I researched how AI could assist individuals with dementia. At the time, AI in healthcare was a nascent field, but its potential was undeniable. I started developing prompting technologies, leveraging algorithms to guide individuals through daily tasks such as washing their hands or dressing. These technologies aimed to provide gentle, context-aware cues tailored to the specific needs and capabilities of the user.

One of my earliest breakthroughs involved using computer vision and machine learning to recognize when a user struggled with a task. By analyzing patterns of movement and behavior, the system could intervene with helpful prompts at just the right moment. These technologies were not just about task completion—they were about preserving dignity, fostering independence, and easing the burden on caregivers.

14.4 Building a Career in AgeTech

Over the years, my work in AgeTech expanded to encompass various technologies to support older adults and their caregivers. From smart home systems that detect falls to AI-powered companions that alleviate loneliness, my research has always been guided by the principle of user-centered design. This approach emphasizes the involvement of older adults and their caregivers in the design process, ensuring that our technologies align with their real-world needs.

My role as a professor at the University of Toronto and my leadership in initiatives like AGE-WELL have allowed me to collaborate with brilliant minds from diverse disciplines. These partnerships have fueled innovation and accelerated the development of technologies that enhance the quality of life for older adults. But while my professional achievements are a source of pride, my personal experiences have truly deepened my understanding of the importance of this work.

14.5 Becoming a Caregiver for My Parents

In recent years, I have taken on the role of caregiver for my parents, both of whom are nearing their 90 s and facing cognitive and mobility opportunities and challenges. This experience has been both deeply rewarding and profoundly humbling. It has given me a firsthand perspective on caregiving's physical, emotional, and logistical demands.

One of the greatest opportunities and challenges has been helping my parents maintain their independence while ensuring their safety. Technologies I once studied in a clinical or academic setting, such as fall detection systems and smart home sensors, have become part of my daily life. This dual role as a caregiver and a researcher has provided invaluable insights into the real-world application of AgeTech. It has also reinforced my belief in the importance of designing technologies that are not only effective but also empathetic.

14.6 Facing My Challenges: Aging with a Disability

In 2019, my life took an unexpected turn when I was involved in an accident that resulted in a spinal cord injury. This experience has profoundly shaped my understanding of aging with a disability. As someone who has spent decades researching technologies for older adults, I relied on many solutions to navigate my opportunities and challenges. From assistive devices that support mobility to smart home technologies that enhance accessibility, I have experienced firsthand the transformative power of AgeTech.

This personal journey has also emphasized the gaps in the field. While technology can explore many physical opportunities and challenges, it often fails to address the emotional and psychological aspects of living with a disability. My experience has renewed my commitment to developing holistic solutions that address the full spectrum of needs.

14.7 Lessons Learned Along the Way

Reflecting on my journey, several key lessons stand out:

14.7.1 Empathy is Essential

Empathy has been the cornerstone of my work in AgeTech. Whether designing technologies for older adults with dementia or developing solutions for my parents, understanding the user's perspective has been critical to creating meaningful innovations.

14.7.2 Collaboration Fuels Innovation

The AgeTech field thrives on interdisciplinary collaboration. Every voice, from engineers and healthcare professionals to caregivers and older adults, is vital in shaping technologies that make a difference.

14.7.3 Technology Alone is not Enough

While technology can explore many opportunities and challenges, it must be complemented by policies, education, and community support to achieve lasting impact. This holistic approach is essential for creating age-friendly societies.

14.8 The Future of AgeTech

As the global population ages, the need for AgeTech will only grow. Advances in artificial intelligence, robotics, and wearable devices hold immense promise, but ethical considerations and a commitment to inclusivity must guide them. My vision for the future of AgeTech is that technologies are seamlessly integrated into daily life, empowering older adults to live independently and with dignity.

14.9 Conclusion

My journey into AgeTech has been shaped by serendipity, personal experience, and a deep-seated desire to make a difference. From that fateful conversation with an engineer about his wife's dementia to my current role as a caregiver and someone aging with a disability, each chapter of my life has reinforced the importance of this field. AgeTech is not just about innovation but about creating a better future for older adults, caregivers, and society. For me, this work is both a professional calling and a deeply personal mission.

Should I Go into a Career in AgeTech? 15

15.1 Introduction

I have always been intrigued by how our world adapts to shifting demographics, especially when it comes to aging populations. It seems that almost every day, a new statistic or a fresh headline is pointing out how quickly our societies are graying. Whether it is a feature in a major newspaper or a conversation with a friend whose grandparents need additional support, the reality of global aging is impossible to ignore. In my own life, I have witnessed how older relatives manage their health and maintain independence. Sometimes, they struggle with technology or feel uncertain about how it can help them—while at other times, a simple digital tool brings them renewed independence. Experiences like these have fueled my fascination with how technology can truly transform the landscape of aging.

As the global population ages, economies, healthcare systems, and social structures must evolve in tandem. According to the [2], by the year 2050, more than 1.5 billion people worldwide will be aged 65 and older [2]. That projection, in and of itself, is staggering. This shift means that many of us will either care for older adults, become older adults ourselves, or work in sectors that aim to support the aging process—if not all three. The breadth of the aging population also suggests that new demands will continue to emerge, spanning every area from public policy to personal well-being.

In this context, AgeTech, broadly defined as the intersection between aging and technology, has become one of the most dynamic frontiers for innovation. I remember the first time I heard the term "AgeTech." I found it fascinating that so many different types of technology, from wearable health monitors to socially assistive robots, could be folded into a single category. Since then, I have discovered that AgeTech includes telemedicine platforms, digital therapy tools, and even entire startups that serve older adults' needs.

A. Mihailidis, *The Age of AgeTech*, Synthesis Lectures on Technology and Health,
https://doi.org/10.1007/978-3-032-04886-8_15

As I dug deeper, it became increasingly clear to me that AgeTech is as broad as it is promising.

From a personal standpoint, I am often drawn to fields where I can blend my technical interests with a sense of social impact. That is precisely why AgeTech stands out to me. It offers a glimpse into some of the coolest emerging technologies, like AI-driven diagnostics, robotic caregivers, and innovative software platforms, and addresses the human side of aging, where empathy, community, and respect for older adults remain paramount.

I believe AgeTech is worth serious thought for anyone considering a career path. This sector is more than just a passing trend; it is a robust ecosystem continually fueled by the growing needs of an aging population. Career opportunities abound in research, industry, policy, and entrepreneurship, allowing individuals to match their own skill sets and passions to the area where they can make the most impact. However, it's important to ask: "Is AgeTech the right career path for me?" That is the question this essay attempts to explore in depth. By examining the sector's potential, the roles available, and the specific skills required, my goal is to help you evaluate whether AgeTech aligns with your professional aspirations and personal values.

15.2 Why AgeTech Matters?

15.2.1 Demographic Trends

One of the primary reasons AgeTech has become so vital lies in the dramatic demographic changes happening around the world. The World Health Organization [3] notes that by 2030, one in six people worldwide will be over 60 years old. Furthermore, for the first time in history, older adults will soon outnumber children under five. That statistic alone stops me in my tracks, because it signals a monumental shift in how our societies function. When I imagine a world where so many people are in their 60, 70 s, and beyond, I see both potential and responsibility: potential for celebrating longer lifespans and responsibility to ensure quality of life at advanced ages.

In conversations with friends and colleagues, the notion of "living longer" often comes up as either a triumph of modern medicine or a challenge for social services (and sometimes both). In my view, these demographic trends are neither purely positive nor purely negative; they simply present a new reality. On one hand, as people live longer, they also contribute more wisdom and experience to communities. On the other hand, healthcare systems can become strained, and the risk of isolation or age-related conditions grows as well.

AgeTech sits at the intersection of these realities. By focusing on how technology can serve older adults, whether that's through better health monitoring, improved connectivity, or more efficient service delivery, we create opportunities to harness the benefits of extended lifespans while mitigating potential downsides. It's a delicate balance, but one that I believe AgeTech professionals are well-positioned to address.

15.2.2 Economic Impact

Economically, the aging population has given rise to what is often called the "silver economy." This term refers to all goods and services targeting older adults, encompassing sectors like healthcare, hospitality, finance, and beyond. Oxford Economics projects that the silver economy will reach a global value of about $15 trillion by 2030. When I first learned about that figure, I remember thinking it was almost too large to grasp. But the more I read and the more I observed real-life examples, the more it made sense.

On a purely economic level, older adults hold significant purchasing power. Additionally, they are an expanding consumer group for products that enhance mobility, health, and daily convenience. AgeTech innovations, such as assistive devices, telehealth platforms, and specialized financial technologies, are key drivers in this market. Companies that identify opportunities to make older adults' lives easier often find that this is not merely a niche demographic—it is a large and growing market segment that can be both profitable and socially valuable.

If you are someone who wants to align your career with an economically robust field, AgeTech stands out as an area primed for growth. Yet, it is not purely about chasing market trends. There is a compelling philanthropic element too: the more AgeTech grows, the more we can help older individuals lead richer, healthier lives. That duality of financial promise and social impact resonates deeply with me, reinforcing my commitment to the field.

15.2.3 Societal Benefits

Beyond sheer economics, AgeTech has a profound role in enhancing the quality of life for older adults. When I talk to people outside the AgeTech space, I like to illustrate this point by sharing anecdotes about specific technologies. For instance, there are robotic caregivers like Japan's Robear, designed to reduce the physical strain on human caregivers while assisting older adults with tasks such as mobility. Systems like that not only address labor shortages in caregiver roles but also provide a safer environment for both the older adult and the caregiver.

In Canada, telehealth platforms such as Maple enable remote consultations, so people living in underserved regions can still have access to a range of healthcare providers. With telehealth, we see a tangible reduction in the barriers many older adults face when traveling to clinics or hospitals. This brings to mind a story of a family friend who lives in a rural area and found it very difficult to travel for regular check-ups. After adopting a telemedicine approach, not only did she receive more consistent care, but she also felt more empowered in managing her condition.

From my perspective, these societal benefits encapsulate the heart of AgeTech. The technology is not designed just to be flashy or to make money, it's there to tackle real-life issues like social isolation, difficulty accessing healthcare, and the desire for greater independence. It amazes me how a seemingly simple piece of tech, such as an app that reminds you to take your medication, can significantly improve daily life and mental well-being. I believe that is why AgeTech draws individuals who crave meaning in their professional paths. It is an arena where technological innovation meets human compassion.

15.3 Careers in AgeTech

15.3.1 Research Careers in AgeTech

Whenever I have spoken with academic mentors or colleagues involved in AgeTech research, I've noticed how passionately they discuss the intellectual puzzle posed by aging. Research forms the foundation of every AgeTech breakthrough, whether it involves discovering new human–computer interaction models for seniors with visual impairments, exploring how robotics might support older adults with physical tasks, or investigating geroscience to understand the biological roots of aging.

15.3.1.1 Skills and Qualifications for Research

If you are drawn to research, you likely already relish the thought of deep dives into scientific literature and intricate data analysis. Researchers often hold advanced degrees in areas such as engineering, computer science, gerontology, or health sciences. For me, the interdisciplinary nature of AgeTech research stands out as particularly appealing. I love the idea that an engineer might collaborate with a geriatrician, a psychologist, and a sociologist—all in service of creating solutions for older adults.

Besides having domain-specific expertise, a researcher in this field also needs proficiency in research methodologies and data analysis. That might involve running randomized controlled trials for a new fall detection system or employing qualitative methods like focus groups to assess how older adults interact with prototype software. A friend of mine who works on usability studies mentioned that her biggest surprise was how important empathy became in her data collection; she realized she had to put herself in the participants' shoes to ask the right questions and interpret results accurately.

15.3.1.2 Opportunities and Examples

Some standout examples in this realm are Canada's AGE-WELL Network and the U.S. National Institute on Aging (NIA). AGE-WELL funds interdisciplinary research that runs the gamut from cognitive health tools to smart home systems, making it a hub for anyone wanting to blend technology with an understanding of seniors' lived experiences. Meanwhile, the NIA invests in research on robotics, AI, and telemedicine for older adults—providing an array of grants and fellowships that can support innovative projects. Whenever I come across new publications from these entities, I am struck by how diverse and impactful their work is, covering everything from the molecular basis of diseases to large-scale social interventions.

15.3.1.3 Pros and Cons

Of course, a research career in AgeTech has its upsides and downsides. On the positive side, you can expect engaging intellectual challenges and the chance to achieve societal impact. I have seen how gratifying it can be to publish academic papers that genuinely move the needle on how older adults are cared for or supported. There is also a certain level of academic recognition that comes with contributing to a niche yet rapidly growing field.

On the flip side, funding competition can be fierce. Grants often require a detailed proposal that clearly articulates both scientific rigor and practical relevance. Research timelines can also be quite long, especially if you are testing new devices in real-world settings or dealing with human participants. Additionally, researchers may sometimes feel removed from end users because much of their work occurs in laboratories or through theoretical frameworks. Bridging that gap is a challenge, but, in my opinion, it's a worthwhile one.

15.3.2 Industry Careers in AgeTech

The private sector is where much of the AgeTech innovation quickly translates into tangible products and services. From what I have observed, industry careers in AgeTech offer a faster-paced environment compared to academic research, with a strong emphasis on product timelines, market validation, and revenue growth. That doesn't make them less impactful; in fact, this fast pace can bring real solutions to real people more swiftly.

15.3.2.1 Key Roles

Within the AgeTech industry, there is a range of roles to consider. Product development engineers might focus on designing and prototyping hardware—think wearable health monitors or mobility aids that help older adults walk more stably. AI specialists, meanwhile, develop algorithms that underpin telehealth platforms or cognitive health tools. A close friend of mine works as a data scientist on a fall detection system; she spends her days optimizing machine learning models to reduce false alarms while ensuring seniors are protected.

Marketing professionals also play a key part in connecting innovative products to the end user. They need to understand not just the technical features of a solution but also the emotional factors that influence whether older adults (and their families) feel comfortable adopting it. Crafting marketing campaigns that respect older adults' dignity and avoid condescension can be a nuanced process.

15.3.2.2 Examples of Companies

When it comes to industry players, some examples jump out immediately. In Canada, PointClickCare leads in providing software solutions for long-term care facilities, revolutionizing how data is shared among nurses, administrators, and families. In Israel, Intuition Robotics has gained attention for ElliQ, an AI companion that fosters social interaction and helps older adults navigate daily tasks. Over in the United States, Google Health invests in applying AI to healthcare, an area that inevitably intersects with aging, given the prevalence of chronic conditions in older populations. Observing these companies, I am often impressed by the extent to which they combine cutting-edge technology with practical, user-centered design.

15.3.2.3 Pros and Cons

One appeal of industry careers is the competitive salary and the possibility for rapid career growth. If you are driven by innovation and love seeing ideas materialize fast, the private sector can be deeply rewarding. Another advantage is the chance to collaborate across multiple divisions, engineering, product management, marketing, and more, to bring a solution from concept to market.

Nevertheless, industry jobs can sometimes feel removed from the direct daily experiences of older adults, especially if market decisions prioritize profitability over user feedback. The profit-driven atmosphere can also lead to intense workloads, tight deadlines, and a pressure to justify return on investment. However, it's entirely possible to find organizations that place a premium on ethical development and user-centered design, striking a balance between business goals and genuine care for seniors.

15.3.3 Policy Careers in AgeTech

For me, policy careers in AgeTech represent a different yet equally critical angle. Instead of designing devices or running experiments, policymakers create frameworks that determine how AgeTech is funded, regulated, and distributed. If we look at all the innovative products in this space, none would succeed without supportive regulatory structures and well-thought-out funding pathways.

15.3.3.1 Skills and Qualifications

Professionals who excel in policy roles typically have backgrounds in public health, law, political science, or economics. They also have in-depth familiarity with aging-related policies, healthcare systems, and technology regulations. Advocacy and communication skills are paramount. I once attended a conference where a policymaker spoke passionately about why she championed the expansion of telemedicine licenses across states. Her ability to articulate the issue clearly, and rally different stakeholders around a common goal, showed me how critical communication is in shaping AgeTech policy.

15.3.3.2 Key Responsibilities

Regulatory oversight is one of the biggest tasks. AgeTech products often fall under medical device or health IT regulations, requiring that policymakers ensure these products meet safety and accessibility standards. Funding allocation is another element; developing grants or incentives to spur AgeTech research and entrepreneurship can profoundly influence which ideas get to market. Lastly, policy development often ties into creating age-friendly initiatives that incorporate technology into public services—whether that means equipping libraries with telehealth kiosks or launching city-wide fall prevention programs.

15.3.3.3 Examples of Initiatives

On a global scale, Japan's Robot Care Initiative stands out. The Japanese government invests heavily in integrating robotic caregivers into long-term care facilities recognizing that robots can help mitigate caregiver shortages in an aging society. Similarly, the European Union's Horizon Europe Framework dedicates funds to aging and technology research, demonstrating a firm commitment to turning innovative ideas into reality.

15.3.3.4 Pros and Cons

Engaging in policy allows you to see your efforts ripple through entire systems, potentially impacting thousands or millions of people. That level of societal influence is deeply inspiring, at least for me. However, policy moves at a different pace than industry or research—bureaucratic processes can slow implementation. Additionally, policymakers often have limited direct involvement in developing technology solutions. That said, for those who thrive on systemic change and stakeholder engagement, a policy career in AgeTech can be exceptionally rewarding.

15.3.4 Startups and Entrepreneurship in AgeTech

Entrepreneurship in AgeTech is perhaps the most high-risk, high-reward path. I find it thrilling to watch startups launch innovative products—some that we might have once only dreamed of in sci-fi films. Founders and employees at these startups need creativity, resilience, and a willingness to adapt quickly.

15.3.4.1 Key Success Factors

A deep understanding of older adults' needs and preferences is absolutely crucial. I have seen promising ideas fall flat because the founders did not take time to speak with seniors or caregivers, leading to products that solved hypothetical issues rather than real pain points. Securing funding through venture capital, grants, or impact investors is another major challenge, and it requires a compelling pitch that convincingly lays out the product's market potential and its social value.

Additionally, assembling a multidisciplinary team can make or break a startup. If you have tech gurus but no marketing or healthcare expertise, you will struggle to navigate regulatory approvals or user engagement. Conversely, if you lack technical experts, your product might remain stuck at the concept stage. I always emphasize that collaborative spirit is paramount in any AgeTech venture—no single person can cover all the bases given the complexity of both aging and technology.

15.3.4.2 Examples of Startups

Some notable startups have captured international attention. For example, Kinova Robotics (Canada) has developed robotic arms that assist individuals with mobility impairments, giving them greater independence in daily tasks. Honor Technology (United States) has modernized caregiving by creating a platform that easily connects families with caregivers. Meanwhile, ActiveProtective (United States) designed a wearable airbag specifically to protect against hip fractures during falls—a challenge that can lead to serious complications for many older adults. Observing how these startups iterate on their products, incorporate user feedback, and scale their operations offers valuable lessons in both entrepreneurship and human-centered design.

15.3.4.3 Pros and Cons

On the positive side, entrepreneurs can rapidly innovate and pivot as they learn from real-world experiences. There is also a sense of personal fulfillment when your idea directly improves someone's life, and you are at the forefront of shaping the AgeTech industry. However, startups face considerable uncertainty, from funding instability to fierce competition. The workload is often intense, and there is always the possibility that a great concept will fail to catch on. Despite these risks, many founders believe the potential impact on older adults and society is worth the gamble.

15.4 Skills Needed for a Career in AgeTech

Regardless of where you land—in research, industry, policy, or entrepreneurship—AgeTech demands a particular constellation of skills. On the technical side, proficiency in AI, robotics, the Internet of Things (IoT), and data analytics is increasingly important. I have seen job postings that ask for a blend of software engineering expertise and a background in healthcare data, reflecting the multidisciplinary reality of this field.

At the same time, empathy and communication skills are non-negotiable. Since you're creating technology for older adults, you need to understand their lived experiences, challenges, and desires. Personally, I have noticed that when product teams or researchers lack empathy, the outcomes tend to feel disconnected from real users. Interdisciplinary collaboration also tops the list of essential skills. Few AgeTech solutions can be effectively built by a single type of expert. Instead, teams often consist of healthcare providers, engineers, social scientists, designers, and business professionals all working in unison.

Finally, adaptability stands out as a make-or-break quality. The landscape of AgeTech changes rapidly due to technological advances, policy updates, and shifting consumer preferences. If you thrive in an environment where you are constantly learning and pivoting, AgeTech might be a perfect fit. If you prefer a slower, more predictable pace, it might be challenging to keep up with the field's demands.

15.5 Global Perspectives on AgeTech Careers

15.5.1 Canada

In reflecting on Canada's AgeTech ecosystem, I am struck by how strongly it emphasizes collaboration and user-centered approaches. The AGE-WELL network epitomizes this philosophy by bringing together researchers, older adults, and industry partners to tackle challenges ranging from cognitive health to smart home technologies. I recall attending a webinar hosted by AGE-WELL, where seniors themselves presented alongside tech developers—a testament to how deeply Canada values participatory research.

Canada's government also plays a role in shaping AgeTech careers. Programs like the New Horizons for Seniors Initiative [1] fund projects that integrate technology into community-based services. If you live in Canada or are considering relocating there, you will find a supportive environment for AgeTech, replete with grants, networks, and academic-industry collaborations. However, there is still a long way to go.

15.5.2 Japan

Japan's status as one of the fastest aging countries in the world makes it a hotbed for AgeTech innovation, particularly in robotics and AI. Panasonic, SoftBank Robotics, and other major Japanese firms have rolled out solutions such as robotic exoskeletons and socially assistive robots. In addition, the government invests heavily in programs to pilot robotic caregivers in long-term care facilities. The synergy between industry and government creates numerous career pathways, from research labs that push the boundaries of robotics to policy teams that streamline adoption in public healthcare settings.

Given Japan's cultural respect for older adults, working in AgeTech there can also be quite fulfilling on a personal level. Each time I have read about new robotic solutions or remote monitoring systems pioneered in Japan, I am struck by how seamlessly they aim to blend technology with daily life, respecting cultural nuances around privacy and family involvement.

15.5.3 United States

The United States offers a fertile ground for AgeTech entrepreneurship, fueled by a robust venture capital ecosystem and strong tech infrastructure. Silicon Valley and other innovation hubs continually spawn new startups—some of which focus explicitly on aging-related products. At the same time, established tech giants like Google and Amazon have ventured into healthcare, naturally extending into solutions for older adults.

Government agencies, like the National Institute on Aging (NIA), fund a significant amount of AgeTech-related research, including studies on telemedicine, cognitive health, and innovative caregiving solutions. Over the years, I have seen many success stories where initial NIA funding helped a technology progress from a prototype to a viable commercial product. For those seeking diverse career options, from fast-paced startups to well-funded labs, the U.S. landscape is as broad as it is vibrant.

15.5.4 Europe

Europe is distinctive for its emphasis on ethical frameworks and inclusivity in AgeTech development. The European Union's Horizon Europe Framework supports interdisciplinary research into aging, aiming to integrate advanced technologies in ways that align with European values on privacy, data protection, and social equity. I find that European AgeTech projects often build strong human-centered design principles from the ground up. Many solutions are co-created with older adults to ensure that even advanced AI systems maintain transparency and fairness.

Because of the EU's regulatory oversight, particularly the General Data Protection Regulation (GDPR), AgeTech innovators must be more mindful of privacy concerns. This can slow down the development process but also leads to robust, ethically sound solutions. Working in Europe could be particularly appealing if you're drawn to policy-driven approaches, multi-country collaborations, and a landscape where ethical considerations hold significant weight.

15.6 Challenges in AgeTech Careers

Despite the excitement and promise, AgeTech careers come with their fair share of challenges, or, as I sometimes like to frame them, opportunities for creative problem-solving. One notable challenge is navigating regulatory barriers. Many AgeTech products, especially those labeled as medical devices, must undergo rigorous approval processes, which can be lengthy and costly. In some regions, the labyrinth of approvals can delay the time it takes to bring a life-changing product to market.

Funding competition is another reality. Although there are grants, venture capital, and impact investors who focus on older adult markets, the growing popularity of AgeTech also means more individuals and companies are vying for the same resources. This pressure can inspire innovation, but it can also be discouraging if you are new to the field or working on a niche technology.

User resistance is a more subtle issue. As someone who has spent time introducing new apps and devices to my older relatives, I have seen firsthand how skepticism can emerge. Some older adults worry about privacy, feel intimidated by digital interfaces, or dismiss a product as something they do not really need. Overcoming these adoption barriers often requires thoughtful design choices, clear communication, and ongoing support. I think it is particularly important for AgeTech professionals to listen to constructive criticism from seniors themselves, adjusting solutions to lower the intimidation factor.

15.7 Future Directions in AgeTech

Looking forward, the AgeTech sector is on the cusp of even more exponential growth—spurred by breakthroughs in AI, robotics, telemedicine, and beyond. One compelling trend is personalized healthcare, where AI-driven tools tailor care plans to each individual. For instance, algorithms can analyze data from wearables, medical records, and environmental sensors to predict the onset of certain conditions or to recommend lifestyle changes. I find this aspect of AgeTech exciting because it aligns well with an era in which we seek precision in healthcare, moving away from one-size-fits-all strategies.

Aging-in-place technologies also continue to evolve. Imagine a home equipped with advanced sensors that detect changes in gait or posture, automatically adjusting the lighting to prevent falls or sending alerts if something seems off. As society aims to help older adults remain in their homes longer, such tech solutions become indispensable. Personally, I am looking forward to seeing more user-friendly and discreet systems that blend seamlessly into everyday life.

Lastly, sustainability is emerging as a concern. As we scale up AgeTech solutions, we need to be mindful of energy consumption, waste management for discarded electronics, and the overall carbon footprint. To me, sustainability in AgeTech is not just an environmental imperative; it is also about ensuring that older generations can continue enjoying the beauty of the world around them. If we are truly committed to a better future for seniors, it includes preserving the planet they live on.

15.8 Conclusion

Deciding on a career in AgeTech is a significant choice—one that intersects technology, healthcare, empathy, and business. From my personal perspective, the field's combination of social impact and rapid innovation is what makes it so compelling. As populations age, the demand for thoughtful, evidence-based solutions will only rise, positioning AgeTech professionals at the forefront of one of the twenty-first century's most urgent domains.

Whether you are drawn to the meticulous world of academic research, the dynamic environment of industry, the systemic influence of policy, or the creative rush of entrepreneurship, there is a place for you in AgeTech. Along the way, you will likely need to cultivate a unique blend of skills: technical prowess, emotional intelligence, interdisciplinary teamwork, and the ability to adapt quickly in a fast-evolving field.

Reflecting on everything we have covered, from demographic shifts to personal experiences with older relatives, it's clear that AgeTech isn't just about new gadgets or software. It is about reshaping how we think about aging altogether, forging societies where longevity is celebrated and supported. Personally, whenever I meet people dedicating their careers to creating technologies that help older adults remain active, independent, and socially connected, I feel a deep sense of hope. It reminds me that with the right blend of innovation and empathy, we can tackle even the most daunting challenges presented by an aging world.

If you find that prospect energizing, if the idea of bridging high-tech innovation with the deeply human aspects of aging speaks to you, then AgeTech might indeed be your calling. It's a path filled with complex questions but also profound rewards. In my mind, embarking on a career in AgeTech means embracing the opportunity to contribute to a global conversation about how we care for each other across generations. And that, in my view, is one of the most meaningful pursuits anyone could undertake.

References

1. Government of Canada. (2024). About the New Horizons for Seniors Program. Retrieved from https://www.canada.ca/en/employment-social-development/programs/new-horizons-seniors.html
2. United Nations. (2024). World Population Prospects 2024. Retrieved from https://www.un.org/development/desa/pd/world-population-prospects-2024
3. WHO. (2015). *World report on aging and health.* Retrieved from Geneva: https://www.who.int/publications/i/item/9789241565042/

The Future of AgeTech: Bridging Science Fiction and Reality

16

16.1 Introduction

I have always been fascinated by the ways in which technology can reshape our lived experiences, especially when it comes to aging. Over the past two-plus decades, I have witnessed a growing momentum in the field of AgeTech, a term I use to describe all technologies designed to enhance the lives of older adults. As someone whose grandparents relied on a patchwork of medical gadgets and caregiving support, I have seen firsthand how even small technological innovations can go a long way in promoting independence and comfort in later life. Yet as I learned more about the emerging landscape of AgeTech, I realized that many innovators are not just trying to meet basic needs or reduce certain risks, they are dreaming of a future that reimagines the entire concept of aging.

According to [1], by 2050, the global population of individuals aged 60 and older will exceed two billion people, which is more than 20% of the world's total population [1]. That demographic projection is staggering. And while some see a crisis brewing, I see an opportunity for humanity to rewrite the narrative of what it means to get older. Indeed, as AgeTech grows more ambitious, it could challenge us to view aging as less of a decline and more of a continued adventure—one supported by smart devices, robotics, AI, and perhaps concepts that border on science fiction.

What if, for instance, we could engineer new organs on demand or even preserve consciousness in a digital format? Ideas like these may sound like wild speculation straight out of Star Trek or Black Mirror. But these science fiction visions have a knack for sparking the real-world innovation that ends up defining our future. Not too long ago, telepresence robots and wearable health trackers sounded equally futuristic, yet they are now widely available. My personal belief is that, while real-world constraints will certainly shape what is possible in AgeTech, it is essential to entertain these out-of-the-box

A. Mihailidis, *The Age of AgeTech*, Synthesis Lectures on Technology and Health,
https://doi.org/10.1007/978-3-032-04886-8_16

notions. The moment we stop imagining radical possibilities, we risk stalling progress that could genuinely transform countless lives.

Of course, existing AgeTech already addresses pressing needs: better healthcare access, improved mobility solutions, and technologies that support independence at home. But beyond these current priorities, the future of AgeTech may take us into truly novel territory, where everything from robotic ecosystems to digital immortality expands the possibilities of human life in ways we could hardly have envisioned just a decade ago. In what follows, we will delve into some of these imaginative, at times whimsical, and occasionally provocative ideas. From the influence of science fiction tropes on emerging technology to the ethical and societal considerations these innovations raise, I will explore how we might change the way we think about aging itself.

16.2 The Science Fiction of Aging

Many of us grew up watching science fiction shows where advanced civilizations used futuristic devices to maintain near-perfect health, travel between galaxies, or even cheat death itself. At that time, those stories felt thrilling but remote, implausible fantasies that gave us cautionary tales or hope for a far-off future. Yet over the years, I have come to appreciate how science fiction often serves as a "design fiction," an imaginative springboard that compels us to rethink the boundaries of human potential. Whether it is Star Trek's tricorder diagnosing diseases in seconds or Altered Carbon's cortical stacks housing human consciousness, these fictional concepts challenge our assumptions about aging, mortality, and identity.

In the realm of AgeTech, three science fiction themes stand out for their profound possibilities:

- **Digital Immortality**—The idea of digitizing human consciousness is both exhilarating and unsettling. In countless sci-fi narratives, people upload their minds into computers, effectively preserving their personalities beyond the demise of their physical bodies. I cannot help but wonder how that would feel—would you still be "you" if your memories were floating around in an AI system? For many, this crosses ethical and existential lines, but it also ignites a spark of curiosity about whether we can transcend the very concept of a finite human lifespan.
- **Biotechnological Enhancements**—Regenerating or replacing aging organs and tissues is another motif that crops up in everything from futuristic TV shows to classic novels. Imagine never having to worry about heart disease or kidney failure because an advanced 3D bioprinter could generate a perfect replacement organ. For me, this resonates on a personal level.

- **Virtual Escapism**—In many works of speculative fiction, older adults are rarely the protagonists. But what if VR technology were so immersive that seniors could seamlessly enter and navigate virtual realms? Rather than feeling the aches of arthritis or the burdens of reduced mobility, they might spend their days exploring realms akin to those in Ready Player One. The potential for social connection and lifelong learning in these digital spaces is enormous, but again, these visions carry their own ethical and psychological complexities.

Altogether, these three themes anchor the science-fiction imagination around aging, showcasing how advanced technology might radically reshape human life. They also point to a fundamental question: As AgeTech evolves, will it only meet baseline needs, or will it eventually embrace some of these more radical possibilities?

16.3 Redefining Independence Through Robotics

16.3.1 Robotic Ecosystems

One of the most immediate ways we see science fiction's influence on real-world AgeTech is in the realm of robotics. When I think of older adults contending with daily challenges, climbing stairs, cooking meals, or remembering to take medication, I envision a future where robotic support could be as ubiquitous as smartphones are today. Rather than having a single robot that performs a narrow function, I imagine entire robotic ecosystems integrated throughout a home, each specialized in a particular task but interconnected via a central AI system.

For instance, consider Autonomous Care Pods. I visualize them as small, wheeled machines, perhaps waist-high, equipped with sensors and a friendly user interface. They could follow individuals around, offering help with mobility or fetching items when needed. If you have seen the animated series The Jetsons, you might recall Rosie the Robot scuttling around the house, tidying up and cracking jokes. Although that concept was once pure fantasy, we now have the technology to create advanced prototypes of such pods. These robots could track a user's movements, predict if they are about to stand up from a chair (and thus be ready to stabilize them), and even remind them to take their medication. By analyzing behavioral patterns over time, the robotic ecosystem would learn the individual's habits, preferences, and emotional states. It is almost as though you have an ever-present companion who also knows precisely how to support you physically, cognitively, and emotionally.

16.3.2 Robotic Companions with Emotions

Emotional AI takes this concept a step further. Imagine if these robots not only fetched your slippers but also sensed shifts in your mood and engaged you in conversation when you seemed lonely. In the same way you might vent to a friend or rely on them for emotional support, these devices could hold "empathetic" dialogues. Thanks to improvements in natural language processing and machine learning, we can conceive of robots that interpret facial expressions, tone of voice, and even subtle gestures.

A friend of mine who works in robotics research shared a personal experience with a prototype companion robot that was designed for older adults dealing with mild cognitive impairment. She was surprised at how quickly the participants seemed to bond with the robot, trusting it with their stories and daily frustrations. While some critics worry about emotional AI fostering artificial relationships, there is no denying the potential comfort such devices can bring, particularly to older adults who may feel isolated.

Science fiction has plenty of examples that mirror these ideas: C-3PO from Star Wars or Data from Star Trek are synthetic beings that interact almost seamlessly with humans. Their roles in these fictional universes suggest a future where robots can be more than just tools; they can be friends, helpers, and even guides.

16.4 AI-Driven Healthcare Revolution

16.4.1 Predictive Health Systems

Over the past decade, the healthcare sector has increasingly turned to AI to analyze patient data and detect early warning signs of disease. I personally find AI-driven predictive healthcare fascinating because it inverts the typical reactive model of medicine. Rather than waiting for an older adult to fall ill, these algorithms anticipate medical issues by combing through genetic data, wearables, and environmental factors. The idea is to catch health threats in their earliest stages, potentially increasing the success rate of interventions.

A particularly intriguing concept is the AI "Doctor-in-a-Watch." Envision a sleek wristband that continuously monitors vital signs, such as blood pressure, heart rate, blood sugar levels, and hormonal fluctuations. The device might detect early symptoms of a cold or stress-induced anxiety, automatically alerting you or your healthcare provider. If you are already on medication, it could release microdoses of critical drugs when needed, significantly reducing hospital visits and complications. A family member of mine with diabetes would find such technology life changing, as it could automatically regulate insulin doses, especially overnight when hypoglycemia can be a serious risk.

16.4.2 Digital Twins for Personalized Care

The concept of digital twins, that is virtual replicas of physical objects, has proven remarkably effective in industries like manufacturing and aerospace, so I am excited about its potential in healthcare. Imagine having a digital twin of your body created from your health data. Medical professionals could run simulations to see how you might respond to a new medication or surgical procedure, all before ever touching a scalpel. This approach could greatly reduce the trial-and-error nature of healthcare and lead to highly customized treatments.

For older adults, such simulations might be especially beneficial given that they often have multiple chronic conditions and take numerous medications. A digital twin could minimize dangerous drug interactions by pre-testing them in a virtual environment. I recall an example of someone who was juggling three medications for hypertension, diabetes, and arthritis simultaneously. They had to switch regimens several times before finding a balance with minimal side effects. A digital twin could have streamlined that process significantly.

Science fiction has prefigured such capabilities in works like Star Trek, where the tricorder instantly analyzes a patient's physiology. While we are not quite at the point of scanning someone in seconds, the development of digital twins indicates we are moving closer to that kind of rapid, predictive, and highly targeted medical care.

16.5 Immersive Virtual Realities for Aging

16.5.1 Escape Aging Through Virtual Worlds

I have often wondered what it would be like if our physical bodies weren't our only means of experiencing life's adventures. In recent years, technologies like virtual reality (VR) and augmented reality (AR) have progressed enough to create incredibly immersive worlds. It is easy to imagine how older adults who face mobility constraints, chronic pain, or loneliness might benefit from "escaping" into a beautiful virtual environment. Think of virtual retirement communities, expansive digital spaces where older adults can socialize, play games, attend concerts, or even take classes, all from the comfort of home.

From my perspective, these virtual worlds could also provide older adults with opportunities to express themselves creatively. Maybe someone confined to a wheelchair in real life could choose to move effortlessly in the VR realm, participating in dance recitals or exploring remote mountaintops. There is an empowering aspect here: physical limitations become less defining, replaced by a world limited only by one's imagination. That said, I am conscious of the risk of over-reliance on virtual escapism. Balancing these immersive digital experiences with meaningful real-world connections remains a challenge the AgeTech community needs to address.

16.5.2 Rehabilitation in the Metaverse

VR also holds great promise for rehabilitation. Currently, physical therapy sessions can be painful or repetitive, making adherence difficult. However, if a therapist could guide you through a VR environment, say, a serene beach setting or a simulation of a city street, your exercises might feel less tedious. An older adult recovering from a hip fracture might find it more engaging and motivational to practice walking exercises in a digital garden teeming with butterflies, rather than a sterile hospital gym.

The notion of integrating real-world therapy goals with the gamification aspects of VR could significantly improve outcomes. A friend of mine, who is a physiotherapist, has been looking into early-stage VR rehabilitation programs. She says her patients become more motivated and less discouraged when immersed in these digital spaces, as they can visually track their progress in a fun environment.

When I watch shows that are considered "futuristic," I am reminded that entire virtual universes could be constructed to meet not just entertainment needs but also physical, mental, and social ones. If leveraged responsibly, VR could act as a powerful aid in eldercare, giving older adults a tangible reason to look forward to daily exercise and to remain socially connected despite physical limitations.

16.6 Biotechnological Advancements and Longevity

16.6.1 Regenerative Medicine

When I think about the older adults in my life, some with chronic conditions or degenerative diseases, I cannot help but be intrigued by breakthroughs in regenerative medicine. Whether it is using stem cells to repair damaged tissues or employing 3D bioprinters to create new organs, the potential to reshape the aging process is immense. The real-world progress in this area has been astonishing. Scientists have already printed functional tissues for research purposes, and organ bioprinting for transplantation is on the horizon.

Imagine a future where needing a new kidney or heart no longer hinges on finding a matching donor; instead, a specialized clinic bioprints a genetically matched organ just for you. The moral and ethical implications run deep, particularly concerns about who gets access to such procedures and at what cost, but it's undeniable that this could alleviate immense suffering. In a more personal context, I have known people desperately searching for organ donors for family members. The stress, uncertainty, and wait times can be harrowing. A world in which organ bioprinting is commonplace could radically reduce such anguish.

16.6.2 Genetic Reprogramming

Another tantalizing avenue in AgeTech is genetic reprogramming. Tools like CRISPR allow scientists to edit genes with remarkable precision. Already, there has been talk of targeting specific genes associated with aging, potentially "switching off" the mechanisms that lead to cellular damage or senescence. It is reminiscent of the fictional "fountain of youth," only this time rooted in molecular science rather than myth.

Some researchers are even investigating whether reprogramming certain genes might reverse signs of aging, at least in mice. While it is still early days, the concept suggests that in the near future, we might see therapies capable of adding decades to healthy human life. The personal appeal here is obvious: who would not want to remain active and vital for a longer period, free from the aches and frailties that often define our later years?

That said, I am also a strong believer in reflecting on the societal impacts of such breakthroughs. If these genetic interventions become available, who pays for them, and how would they alter our existing healthcare and social support systems? As excited as I am about the prospect of a longer and healthier life, I recognize that issues of equity, resource allocation, and informed consent will complicate the path forward.

16.7 Consciousness Preservation and Digital Immortality

16.7.1 Mind Uploading

The concept of mind uploading stands out as one of the most radical ideas in AgeTech, a notion that often feels straight out of a dystopian novel. And yet, I have seen serious ethical and scientific discussions on how we might map the human brain's intricate neural connections and "port" that data into a digital substrate. If successful, this would mean that an older adult's memories, personality, and sense of self could persist even after their physical body has reached its end.

From a purely imaginative standpoint, mind uploading is fascinating. What would it be like to speak to a digitized version of a loved one who had passed away, or consult your own "virtual self" from years prior? I recall losing a close grandparent and wishing I had captured more of their stories and advice for posterity. A digitized consciousness could preserve not just anecdotes and insights, but also something akin to their decision-making patterns and emotional responses.

16.7.2 AI Integration

However, mind uploading need not be a mere static recording. AI could enhance it, leading to what some call "augmented selves." In theory, these augmented selves could continue learning, adapting, and evolving, offering future generations guidance. Imagine an older adult who had built decades of expertise in entrepreneurship or medicine—once digitized, they might be able to mentor people for centuries. This is where science fiction and reality converge in a way that feels both exhilarating and unnerving, raising questions about identity, personhood, and mortality.

Altered Carbon depicts a version of this idea by storing human consciousness in portable "stacks." And while we are nowhere near commercializing consciousness in that manner, the principle—transferring the essence of a person across different bodies or systems, points to a possible future. How we navigate the ethical quagmire remains unclear: Should everyone have the right to preserve their consciousness? Would we limit these technologies to a privileged few? And how would society handle conflicts between a digitized self and its physical counterpart?

Even as I marvel at the possibilities, I feel discomfort. There is a sacredness to our finite lifespans, a narrative arc that shapes our relationships and motivations. Yet the potential to live on, even in digital form, could be deeply reassuring for people who fear oblivion or who yearn to maintain connections with loved ones. Balancing these perspectives will be a defining challenge in the AgeTech of tomorrow.

16.8 AgeTech in Space Exploration

16.8.1 Older Astronauts

It might seem like a leap to connect space exploration with aging, but consider the possibilities if AgeTech solutions become advanced enough to make space travel viable for older adults. If you have followed NASA missions, you know that the physical demands on astronauts are immense. Yet with the right exoskeletons, advanced health monitoring, and regenerative medical therapies, seniors might participate in space missions without the severe risks that typically come with advanced age.

I am particularly drawn to the idea that older adults, with their accumulated wisdom and life experience, could contribute a valuable perspective on interplanetary or interstellar missions. Perhaps exoskeletons adapted for low-gravity settings would help them move around a space station or lunar base. AI-driven healthcare pods on board could continually assess their bone density, muscle mass, and cardiovascular function—areas especially crucial for older individuals in microgravity environments. Rather than viewing age as a limiting factor, we could see it as an asset in a cosmic context.

16.8.2 Interstellar Retirement Communities

Going a step further, some visionaries speculate about orbital assisted living stations, effectively turning space tourism into a form of cosmic retirement. On one hand, this sounds utterly fantastical. But as commercial entities like SpaceX and Blue Origin push the boundaries of private spaceflight, who is to say that specialized senior living environments in low Earth orbit are impossible? Living in space could actually offer certain physiological advantages if combined with advanced medical care. Zero gravity might relieve stress on arthritic joints, for instance.

Of course, this raises considerable questions around cost and feasibility. Space travel remains prohibitively expensive, so an interstellar retirement community might only be accessible to the very wealthy—at least initially. There are also countless logistics to work out, from radiation shielding to mental health considerations. Yet the idea resonates with me because it breaks down the assumptions about who can or should explore the frontier of outer space. If we harness AgeTech effectively, we could find ways to involve older adults in humanity's next great adventure.

It reminds me of the film Interstellar, which tackles cosmic travel as an existential leap for humanity. While that story focuses more on saving humanity than offering retirement solutions, it exemplifies the notion that space exploration does not have to be exclusive to the young and physically prime. With the right AgeTech solutions, age might be just another factor that technology can mitigate, enabling seniors to become cosmic explorers.

16.9 Ethical Considerations for the Future

For all these breathtaking visions, I find it critical to maintain a grounded perspective on the ethical questions. Every leap in AgeTech carries potential pitfalls, and some innovations raise moral quandaries so intense they cannot be answered in a single stroke. Below are some major areas of concern:

- **Privacy and Security**—As robotic ecosystems, AI-driven healthcare, and immersive VR gather more and more data on older adults, how do we ensure that this sensitive information is protected? If a person's daily routines, emotional states, and even genetic information are constantly monitored, the risk of data breaches or misuse looms large. I worry that older adults, particularly those less tech-savvy, could be exploited by unscrupulous entities seeking to profit from their personal data.
- **Access and Equity**—I have repeatedly seen advanced healthcare solutions remain out of reach for lower-income or marginalized communities. If the next frontier of AgeTech includes organ bioprinting, VR rehab, and mind uploading, will these breakthroughs only benefit a wealthy elite? As a society, we have to decide whether AgeTech is a universal right or a luxury good. Personally, I hope we strive toward more inclusive

models that ensure older adults from all backgrounds can access these life-improving (and potentially life-extending) technologies.

- **Human Identity**—Finally, the question that keeps me awake at night is how these innovations might alter our understanding of what it means to be human. If we can "upload" our consciousness or live in VR indefinitely, do we lose something fundamental about the human experience? Or do we gain a new realm of freedom and exploration? Similarly, if we genetically engineer ourselves to postpone aging indefinitely, how does that reshape our notion of a life's journey and natural limitations?

Ethical discussions are not just philosophical musings; they directly influence how we design and deploy technology. That is why I believe ethicists, philosophers, and social scientists must be actively involved in AgeTech research and commercialization. Without their voices, it is too easy to chase after the "cool factor" of these futuristic concepts without grappling with their societal impacts.

16.10 Conclusion

When I step back and look at the landscape we have covered, including robotic helpers with emotional intelligence, predictive AI healthcare, immersive VR worlds, biotechnological breakthroughs, digital immortality, even senior-friendly space travel, I am struck by how quickly these ideas move us beyond just "solving aging problems." Instead, they invite us to redefine aging itself. In so many ways, AgeTech stands at the intersection of science fiction and emerging scientific reality, daring us to imagine a society where the limitations of age are dramatically diminished, and where older adults can experience unprecedented forms of autonomy, health, and even adventure.

Yet I'm also mindful that these visions come with ethical, social, and practical challenges. Technology without thoughtful implementation can exacerbate inequalities or strip away essential human touchpoints. Indeed, the tension between innovation and responsibility is a recurring theme in my conversations with peers in healthcare and tech. As we move forward, it is imperative that we craft policies, regulations, and cultural norms that allow for bold experimentation while protecting human dignity, privacy, and equity.

The ultimate promise of AgeTech is not merely to prolong life, but to enrich it, to provide older adults with fulfilling, meaningful experiences no matter what their chronological age might be. Personally, I find inspiration in the notion that as a global community, we have an opportunity to shape a future where aging is celebrated rather than feared, where each individual's wisdom and identity can continue to evolve, and where technology serves as both a tool and a companion. I find myself growing more hopeful each time I consider these possibilities. If we pair visionary thinking with collective responsibility, there is no telling how far we can go in expanding the horizons of human life.

In the end, what excites me most about these "out-of-the-box" AgeTech ideas is not just the tech itself, but the change in perspective they inspire. Whether it's dreaming up robotic care pods, digital twins, or cosmic retirement communities, these concepts remind us that aging is not a fixed destiny, it is a journey. And thanks to emerging technologies, we may soon have the power to make that journey more empowering, inclusive, and deeply meaningful for everyone who embarks upon it.

Reference

1. United Nations. (2024). World Population Prospects 2024. Retrieved from https://www.un.org/development/desa/pd/world-population-prospects-2024

17 Interdisciplinary Collaboration: Bridging Gaps to Advance AgeTech Innovation

17.1 Introduction

The global population is aging at a pace never before experienced in human history. When I reflect on conversations I have had with colleagues and friends, it often strikes me just how many societies around the world are grappling with the rapid expansion of their older adult populations. The World Health Organization projects that by 2050, there will be around 2 billion individuals aged 60 or older, a sharp increase from 900 million in 2015 [7]. Each time I revisit these numbers, I am reminded of the profound transformations that health systems, social services, and economic structures must undergo to support such a large demographic. Indeed, for many nations, this shift is both an extraordinary opportunity to enrich the lives of older adults and a daunting challenge that calls for systemic change.

Enter AgeTech, the intersection of aging and technology. As someone whose career has spanned both engineering and occupational therapy, I have witnessed firsthand how rapidly emerging technological solutions can bring newfound autonomy and empowerment to older adults. AgeTech offers innovative devices, software platforms, and infrastructure enhancements capable of promoting safety, mobility, and overall quality of life. Yet, based on my personal experiences working on interprofessional teams, I have come to firmly believe that simply inventing gadgets or apps is not enough. What truly propels AgeTech forward is the synergistic collaboration of experts from multiple domains, including engineering, healthcare, social sciences, policy, design, and more.

My career trajectory has given me ample insight into why such interdisciplinary endeavors matter. In one of my earliest roles, I focused almost exclusively on the technical design of assistive devices. Over time, I realized that a purely engineering-driven approach risked overlooking the nuanced needs and emotional experiences of the older adults who would use these tools. It was only after engaging with occupational therapists,

A. Mihailidis, *The Age of AgeTech*, Synthesis Lectures on Technology and Health,
https://doi.org/10.1007/978-3-032-04886-8_17

psychologists, and care providers that I started to understand the daily realities of aging and the human side of technology adoption. Reflecting on these lessons, I now see interdisciplinary collaboration as not just beneficial but essential for AgeTech solutions that truly meet the needs of older adults.

This essay explores the critical importance of cross-disciplinary collaboration in AgeTech. We will examine the inherent complexity of aging challenges, delve into the necessity of bridging engineering and occupational therapy, and highlight contributions from other key fields such as psychology, sociology, and design. We will also look at case studies drawn from my work at the Intelligent Assistive Technology and Systems Lab (IATSL), including the development of an intelligent prompting system called COACH and various ambient assisted living projects. Finally, we will address the challenges, and ultimate rewards, of interdisciplinary work, offer suggestions for the future, and emphasize why collaboration remains vital for true AgeTech innovation. Through these reflections, I hope to underscore a vision where technology not only assists older adults but also respects their dignity and individuality.

17.2 The Necessity of Cross-Disciplinary Efforts in AgeTech

17.2.1 Complexity of Aging Challenges

Aging is anything but a one-dimensional phenomenon. Across my interactions with older adults, I have seen how aging can manifest as physical changes, like decreased mobility or changes in vision, but it can also encompass cognitive shifts, emotional needs (such as the need for social connectedness), and even cultural considerations. It is this multifaceted nature of aging that underscores the complexity we face in designing effective AgeTech solutions. Researchers have long recognized that older adults are not a homogenous group; variations in health status, socioeconomic backgrounds, life experiences, and personal preferences mean that no single solution works for everyone.

As [1] point out, technology must be adaptable, respectful of cultural contexts, and flexible enough to serve a broad spectrum of abilities [1]. For example, a family friend of mine struggled with a supposedly "universal" voice assistant because it did not account for her slight speech impediment. She would often express frustration that the device either misheard her commands or failed to respond in a timely manner. This single example highlights how important it is to view older adults as individuals with varying needs. An engineering team might address the speech-recognition issue by refining algorithms, but real success might also demand insights from occupational therapists, who can offer guidance on how to tailor devices for certain speech patterns, or psychologists who can offer strategies to reduce user frustration. In other words, tackling such a complex challenge requires a team drawn from multiple disciplines.

17.2.1.1 Integration of Engineering and Occupational Therapy

Engineering often takes center stage when it comes to inventing the hardware and software used in AgeTech, yet I have consistently found that it is occupational therapy that brings a unique understanding of human ability, therapeutic processes, and daily living activities. I remember my first interdisciplinary project vividly: our team was designing an assistive robotic arm that could help individuals with limited mobility perform tasks like feeding themselves or reaching for objects. On the technical side, we were laser-focused on the mechanics and control algorithms—making sure the arm moved smoothly and safely. But once occupational therapists entered the discussion, we discovered how much we had overlooked: the emotional impact on users, the best ways to introduce the device in a daily routine, and the subtle cues the device could give to avoid frustration.

This synergy between engineering and occupational therapy reflects the work of [2], who emphasized that technologies must be both functional and person-centered [2]. Without engineering, we would not have the sophisticated tools that allow for precise movement or data collection. Without occupational therapy, we might be producing gadgets that look fantastic in the lab but fall flat when placed into real-world contexts. In essence, bridging these two fields means ensuring that human-centered design is not an afterthought, but a guiding principle from the earliest stages of development.

17.2.2 Role of Other Disciplines

Although engineering and occupational therapy form a powerful foundation, AgeTech solutions demand expertise from multiple additional disciplines. Psychology, for instance, can illuminate why some seniors eagerly adopt new tools while others resist them. Sociologists can uncover how cultural or community norms shape the willingness to rely on technology for daily tasks. Designers ensure that any device, interface, or living environment is not only functional but also aesthetically pleasing and intuitive, which significantly boosts engagement. Meanwhile, economists study cost-effectiveness, helping teams develop solutions that can be scaled and sustained.

Throughout my career, I have had the chance to collaborate with colleagues from each of these domains. Each discipline provides a distinct lens. Psychologists might question how a new device can uphold an older adult's sense of autonomy, while designers could propose simplified user interfaces and gentle color schemes that reduce cognitive load. Such collective insights prevent us from ending up with rigid, one-size-fits-all devices. Instead, they guide the creation of inclusive, flexible solutions with genuine staying power in real-world contexts.

17.3 Case Studies from My Career

17.3.1 Intelligent Assistive Technology and Systems Lab (IATSL)

My experiences at the Intelligent Assistive Technology and Systems Lab (IATSL) at the University of Toronto exemplify the kind of interdisciplinary work that AgeTech demands. The lab is a bustling environment where engineers, computer scientists, occupational therapists, psychologists, and sometimes architects congregate to tackle aging-related challenges. We share a common goal: to create intelligent systems that address the needs of older adults and people with disabilities. Yet each project reveals just how crucial it is for these experts to work in tandem.

Having personally been involved in several IATSL initiatives, I have learned how important it is to begin projects by clarifying what each discipline hopes to contribute. In the early design phase of any project, we hold brainstorming sessions that bring together various professionals. Rather than diving straight into coding or hardware schematics, we first ask basic questions: What real-world issues are we trying to solve for older adults? Which domains of life does it touch—mobility, cognition, communication, or social connection? How can we ensure the final product feels welcoming rather than intimidating? Observing how each person approaches these questions from their disciplinary perspective has shown me that interdisciplinary dialogue is critical for success.

17.3.2 Development of COACH

A standout example of this collaborative spirit is the COACH (Cognitive Orthosis for Assisting with Activities in the Home) system. I have personally found this project to be eye-opening in terms of how an interdisciplinary team can devise technology that significantly improves daily life for older adults with dementia. COACH uses artificial intelligence and computer vision to monitor a person's activity, such as handwashing, and provide prompts if it detects confusion or a stalled step.

Engineers and computer scientists played a pivotal role in developing the underlying algorithms, ensuring that COACH could accurately recognize actions in real time. This involves training machine learning models to distinguish between washing one's hands, drying them, or pausing altogether. Meanwhile, occupational therapists helped define what "appropriate prompts" look like. From a purely technical standpoint, a prompt might just be a beep or text on a screen, but occupational therapy experts knew that individualized prompts, both in timing and tone, could spell the difference between a smooth user experience and a frustrating one. Psychologists similarly contributed their expertise by helping design the user interaction so that it aligned with cognitive processes typical in dementia care. For instance, they guided us on how to make the prompt frequency feel neither rushed nor patronizing.

This project, referenced in [4], exemplifies the deep integration of varied perspectives. The success of COACH lay not merely in its technical sophistication but in its alignment with the behavioral and cognitive needs of its users—something that only a truly interdisciplinary team can accomplish.

17.3.3 Ambient Assisted Living

Another research avenue I have pursued is ambient assisted living (AAL), which seeks to embed smart technologies into living spaces themselves. These environments use sensors, AI, and context-aware computing to gently monitor residents, detect health patterns, and intervene when necessary. According to Sixsmith and Gutman [6], AAL can offer older adults a safer environment by anticipating risks like falls or health deterioration before they become critical.

In developing AAL projects, our teams have included engineers to handle sensor technologies and data processing, healthcare professionals to interpret which metrics are most relevant, and architects or interior designers to ensure the environment remains visually appealing and comfortable. I recall one collaboration where we set up a prototype "smart apartment" outfitted with sensors that monitored everything from movement patterns to air quality. After a few months of real-world testing, we noticed that some older adults felt uneasy about the idea of being "watched" around the clock. Psychologists on our team then investigated how to minimize feelings of surveillance. They suggested more transparent communication about data use, as well as an option for residents to manually pause monitoring if they desired privacy. In essence, it was the interplay of engineering, architecture, psychology, and occupational therapy that made the environment both technologically advanced and acceptable to residents.

17.4 Bridging Engineering and Occupational Therapy

17.4.1 Application of Engineering Principles to Occupational Therapy

From the time I first paired up with occupational therapists on a collaborative project, it became clear that engineering approaches, such as systems design and iterative prototyping, are highly complementary to clinical and therapeutic perspectives. In many ways, occupational therapy is also about problem-solving: identifying a person's functional goals and adapting environments or tasks to help them succeed. Engineering, meanwhile, specializes in designing and testing tools to meet specific requirements. When we fuse these mindsets, we create assistive technologies that are both conceptually sound and pragmatically effective.

For example, if we look at designing a robotic feeding device, engineering might handle sensor placement, mechanical articulation, and control algorithms. However, an occupational therapist would refine how that device integrates into mealtime routines, ensuring it doesn't unintentionally hinder social interaction or erode the sense of agency an older adult might have. By adopting user-centered design, a principle well established in engineering, and pairing it with the empathetic, personalized approach of occupational therapy, we stand a better chance of crafting solutions that older adults embrace wholeheartedly.

17.4.2 Enhancing Rehabilitation Through Technology

Another sphere where engineering and occupational therapy converge is in the development of cutting-edge rehabilitation tools. Virtual reality (VR) systems, for instance, can immerse patients in motivating exercises that replicate real-world scenarios. I once saw a stroke rehabilitation patient practice opening and closing a virtual door, with the VR environment varying the resistance or shape of the doorknob to match the difficulty level. Engineers contributed their software and hardware expertise, ensuring that the VR system accurately tracked arm movements, while occupational therapists created the therapy protocols, deciding what tasks or exercises would yield the best functional improvements.

Similarly, robotic exoskeletons can assist with upper-limb or lower-limb therapy, allowing people to move in ways they could not on their own. In these cases, engineers meticulously refine the torque, speed, and mechanical design, while occupational therapists develop progressive training plans that gradually decrease robotic assistance as the user regains strength. This type of integrated approach aligns with research by Laver et al. [3], which highlights how VR and robotics can significantly enhance rehabilitation outcomes when properly guided by clinical expertise.

17.5 Challenges in Interdisciplinary Collaboration

17.5.1 Communication Barriers

Despite the clear benefits of interdisciplinary work, challenges do arise. One of the most common hurdles I've encountered is communication. Each discipline has its own lexicon and methodologies. Engineers may talk about "signal processing" or "algorithmic complexity," while occupational therapists discuss "client-centered practice" and "functional assessments." The risk of misinterpretation is real, and even well-meaning collaborators can talk past each other without realizing it.

Moreover, different fields sometimes place varying degrees of importance on aspects of a project. An engineer might see a design as "complete" if it functions robustly in test conditions, whereas a healthcare professional could view that same design as incomplete if it fails to account for emotional well-being or practical use in a chaotic real-world environment. These nuances require proactive, patient discussions.

17.5.1.1 Solution: Interdisciplinary Education

My experiences at the University of Toronto underscore the value of interdisciplinary education. We regularly host workshops and courses where students from engineering, healthcare, and social sciences team up to tackle a real-world AgeTech challenge. These students learn each other's jargon, assumptions, and motivations. This immersive approach allows them to cultivate a shared language and fosters mutual respect for each field's contributions. I personally enjoy mentoring such teams because it feels like an investment in the next generation of innovators who already understand the importance of bridging disciplinary divides.

17.5.2 Differing Perspectives and Priorities

Another challenge stems from differing perspectives on what "success" looks like. Engineers might prioritize technical innovation, developing the most sophisticated or efficient device. Healthcare professionals, on the other hand, often care more about patient outcomes, such as functional independence or mental well-being. Meanwhile, funding agencies or startup investors might focus on market viability, wanting to see a clear path to commercialization.

Striking a balance among these varied objectives is no small feat. I have observed projects where the engineering team worked tirelessly to incorporate advanced features, only for healthcare partners to label those features as superfluous if they don't enhance patient quality of life. Similarly, cost constraints can override certain design preferences, forcing the entire team to reassess their priorities.

17.5.2.1 Solution: Collaborative Goal Setting

In my experience, a critical step is to establish common objectives right at the project's outset. During the kickoff phase, we often hold a series of brainstorming and goal-setting sessions, inviting all stakeholders, engineering leads, clinicians, potential users, and sometimes even caregivers, to articulate their aspirations. By the end of these sessions, we craft a unified mission statement that explicitly details both the technical benchmarks (like processing speed or battery life) and the human-centered metrics (like user satisfaction or health outcomes). Regular check-ins ensure that everyone remains aligned, and if priorities shift, the group can discuss how to adapt while preserving core objectives.

17.6 The Impact of Interdisciplinary Collaboration on AgeTech Innovation

17.6.1 Enhanced User-Centered Design

Interdisciplinary collaboration naturally leads to more nuanced, user-centered design processes. With gerontologists, occupational therapists, and psychologists on board, the engineering team gains insights into the lived experiences of older adults, from daily pain points to cultural or generational expectations around technology use. This heightened awareness can dramatically increase user acceptance and satisfaction.

17.6.1.1 Example: Personalized Assistive Technologies

I recall working on a personalized digital assistant for older adults with mild cognitive impairment. Initially, the engineering plan was to build a single interface for all users, but after consulting occupational therapists and psychologists, we realized older adults vary significantly in their memory capacities, comfort with icons or text, and cultural preferences. As a result, we developed a dynamic user interface that could adapt its complexity and language based on user feedback. This personalization only happened because the interdisciplinary team recognized that "one size fits all" would not succeed in the real world.

17.6.2 Improved Technology Adoption

Beyond the design itself, interdisciplinary input is essential for fostering widespread adoption. The best device in the world will not succeed if older adults, caregivers, or healthcare professionals find it confusing or irrelevant. Researchers like [5] have emphasized that a deep understanding of social and psychological factors can significantly improve an innovation's acceptance among seniors.

17.6.2.1 Example: Addressing the Digital Divide

One personal observation is that many older adults remain wary of new devices due to limited digital literacy or negative past experiences. Sociologists and educators have been vital in uncovering these barriers, be they limited Internet access, unfamiliarity with modern gadget interfaces, or fear of data breaches. By working with these experts, we developed training programs and simple user guides that eased anxieties. We also set up community workshops that allowed older adults to experiment with the technology in a supportive environment. Seeing how participants began to warm up to the tools, going from skepticism to enthusiastic daily use, was one of the most rewarding experiences of my career.

17.7 Future Directions

17.7.1 Continued Interdisciplinary Approaches

As AgeTech continues to evolve, new developments in AI, IoT, and biotechnology will open uncharted territories, creating fresh challenges and possibilities. Wearable sensors could track vital signs in real time, while AI algorithms predict emergencies like falls or cardiac events with increasing accuracy. Biotechnology might enable the development of novel implants or even personalized genetic interventions. Each of these emerging areas accentuates the importance of interdisciplinary collaboration. Engineers must work alongside biologists, ethicists, healthcare providers, and social scientists to ensure that these powerful technologies remain safe, accessible, and respectful of users' dignity.

17.7.2 Potential Areas of Growth

- **AI and Machine Learning**: Beyond fall detection, machine learning can generate predictive models for a variety of health events, from depression risk to early signs of dementia. Collaborations among data scientists, clinicians, and psychologists will be indispensable for validating these models in real-world settings.
- **Biotechnology:** As biosensors become more sophisticated, we could see continuous glucose monitoring or early cancer detection integrated into wearable devices. Biotechnologists, engineers, and occupational therapists will need to coordinate on how these sensors affect daily living and how best to interpret data for meaningful interventions.
- **Human–Computer Interaction (HCI)**: For older adults to embrace advanced technologies, interfaces must be intuitive, flexible, and adaptable. This domain demands a synthesis of design principles, cognitive psychology, and engineering. The next wave of AgeTech could include voice interfaces, gesture controls, or even brain-computer interfaces customized for older users.

17.8 My Vision for the Future

17.8.1 Initiatives to Promote Collaboration

Reflecting on my own trajectory, I envision a future where interdisciplinary teams are the norm rather than the exception in AgeTech. To reach that goal, we need proactive steps that encourage collaboration at multiple levels:

- **Interdisciplinary Research Centers**: Establish specialized centers dedicated solely to AgeTech, bringing together experts in engineering, occupational therapy, psychology, sociology, policy, and more. These centers would serve as innovation hubs, providing state-of-the-art facilities and offering ongoing professional development. They might also sponsor hackathons or design sprints where students from diverse fields tackle time-sensitive challenges, fostering a culture of creativity and cooperation.
- **Funding Opportunities**: Grant agencies should explicitly call for collaborative proposals that include multiple disciplines. I have seen how requiring interdisciplinary teams to apply for certain grants fosters synergy, as groups that might not naturally converge realize that partnership is a prerequisite for funding.
- **Policy Development**: Policymakers can support AgeTech by developing frameworks that streamline certification, data privacy, and reimbursement models. It is critical that these frameworks are informed by engineering constraints, clinical requirements, and social considerations all at once. Legislative bodies need diverse advisory panels to ensure a well-rounded approach to AgeTech governance.

17.8.2 Personal Reflection: Emphasizing the Human Element

Whenever I look toward the horizon of AgeTech, I am struck by the dual promise and complexity it holds. We can build extraordinary devices, harness AI to manage chronic conditions, or create environments that support aging in place. But none of those achievements matter if older adults feel alienated or if the solutions fail to address tangible needs. Interdisciplinary collaboration ensures that at every step, from design to deployment, we keep the real beneficiaries, older adults, front and center.

I also maintain a firm belief in the power of empathy. In my personal view, empathy transcends disciplinary boundaries. It is the quality that keeps us all anchored in the user's experience, guiding the sophistication of engineering to serve practical, dignified, and meaningful outcomes. By combining empathy with the technical prowess gained through interdisciplinary collaboration, we elevate AgeTech from a mere collection of gadgets to a movement capable of transforming how we age.

17.9 Conclusion

Interdisciplinary collaboration stands at the heart of truly impactful AgeTech innovation. The demographic shift toward an aging global population is both an urgent call and an exciting opportunity for technology to reshape how older adults live, interact, and receive care. Yet, as my experiences demonstrate, technology alone cannot solve the intricate, multifaceted challenges of aging. Engineers, occupational therapists, psychologists, sociologists, designers, economists, and policymakers each bring irreplaceable insights and skill sets to the table.

By weaving these strands together, we can produce comprehensive solutions that do more than address functional deficits—they empower older adults to maintain independence, social connections, and overall well-being. Projects like COACH illustrate how AI and computer vision can be paired with an occupational therapy framework to support older adults with dementia in daily tasks, while ambient assisted living initiatives show that architects, designers, and healthcare professionals can co-create living environments that are safe yet unobtrusive.

Of course, interdisciplinary collaboration has its pitfalls, such as communication barriers and misaligned priorities. However, proactive strategies like interdisciplinary education and collaborative goal setting can mitigate these challenges. Looking ahead, the AgeTech sector will likely expand into domains such as advanced AI-driven healthcare, biotechnology for continuous monitoring, and new HCI paradigms for older adults. In each of these areas, cross-disciplinary partnerships remain indispensable.

Ultimately, my hope is that we continue to cultivate a culture where no single discipline claims sole ownership of a project, but rather, all contributors work side by side, united by a shared vision. Personally, I find immense satisfaction in seeing how this mindset leads not only to breakthroughs in technology but also to real improvements in the quality of life for older adults. By forging strong interdisciplinary collaborations, and always keeping empathy at the core of our efforts, we stand on the verge of an AgeTech renaissance that could redefine what it means to grow older in the modern world.

References

1. Czaja, S. J., & Lee, C. C. (2007). The impact of aging on access to technology. *Universal Access in the Information Society, 5*(4), 341-349. https://doi.org/10.1007/s10209-006-0060-x
2. Gitlin, L. N. (1995). Why older people accept or reject assistive technology. *Generations: Journal of the American Society on Aging, 19*(1), 41–46.
3. Laver, K. E., Lange, B., George, S., Deutsch, J. E., Saposnik, G., & Crotty, M. (2017). Virtual reality for stroke rehabilitation. *Cochrane Database of Systematic Reviews*(11). https://doi.org/10.1002/14651858.CD008349.pub4
4. Mihailidis, A., Boger, J. N., Craig, T., & Hoey, J. (2008). The COACH prompting system to assist older adults with dementia through handwashing: An efficacy study. *BMC Geriatrics, 8*(1), 28. https://doi.org/10.1186/1471-2318-8-28
5. Peek, S. T. M., Wouters, E. J. M., van Hoof, J., Luijkx, K. G., Boeije, H. R., & Vrijhoef, H. J. M. (2014). Factors influencing acceptance of technology for aging in place: A systematic review. *International Journal of Medical Informatics, 83*(4), 235-248. https://doi.org/10.1016/j.ijmedinf.2014.01.004
6. Sixsmith, A., & Gutman, G. (2013). *Technologies for active aging* (Vol. 9): Springer.
7. WHO. (2015). *World report on aging and health.* Retrieved from Geneva: https://www.who.int/publications/i/item/9789241565042/

18 Global Perspectives: Cultural Considerations in AgeTech

18.1 Introduction

I have always been fascinated by how different cultures perceive and support older adults, especially as the global population continues to age at an unprecedented rate. According to the [13], by 2050, there will be over 2 billion people aged 60 and above, a large proportion of whom will reside in developing countries. In my own work with community groups, I have seen firsthand how families and societies balance tradition with the evolving demands of healthcare for older adults. AgeTech, often called "aging technology," seeks to address these demands by harnessing technological innovations to improve quality of life for seniors. Yet as I have come to realize, a successful AgeTech solution is not solely about cutting-edge devices or software platforms, it also hinges on cultural fit.

Cultural values and norms crucially shape how different societies view aging, family structures, and the role of technology. In some places, there is deep respect for older adults as repositories of wisdom, while in others, aging is perceived less positively. Such attitudes inevitably influence how quickly and effectively new technologies are adopted. Nowhere is this more evident than in the Global South, where cultural frameworks can diverge sharply from Western norms. By understanding these local contexts and integrating cultural insights into AgeTech design, we can create solutions that genuinely resonate with older adults worldwide. In this essay, I will explore the cultural factors affecting AgeTech adoption across various global contexts, focusing especially on examples from Asia, Africa, and Latin America, and propose strategies for developing culturally appropriate technologies.

A. Mihailidis, *The Age of AgeTech*, Synthesis Lectures on Technology and Health,
https://doi.org/10.1007/978-3-032-04886-8_18

18.2 Cultural Influences on AgeTech Adoption

18.2.1 Perceptions of Aging and Older Person Care

Cultural beliefs about aging can significantly shape whether new technological interventions gain acceptance. Some communities celebrate older age as a time of accrued knowledge and social standing, while others see it as a period of diminished capacities. These perceptions directly inform whether families and older adults themselves see AgeTech as helpful or unnecessary. From my experience living in a large, intergenerational household, I have witnessed how family members often take on distinct caregiving roles, something that is very common in collectivist societies. In such environments, technology that appears to reduce personal contact might seem cold or even disrespectful.

- **Collectivist Societies**: In countries across Asia and Africa, family bonds are central to caregiving. Children often feel a moral imperative to provide physical and emotional support for their aging parents. Consequently, an AgeTech device that replaces certain caregiving activities might cause concern or guilt. Rather than perceiving the technology as a beneficial tool, some might view it as a sign of neglect or abandonment.
- **Individualist Societies**: In contrast, older adults in Western contexts may place more emphasis on maintaining personal independence [4]. With fewer cultural obligations for adult children to provide daily care, seniors in these settings may be more receptive to technologies that help them live autonomously, whether that is a wearable health monitor or a home-based telemedicine service.

18.2.2 Technological Literacy and Accessibility

I have often noticed that people's capacity or confidence to engage with digital tools varies widely, sometimes even within the same country. In remote areas, I have met older adults who have never touched a smartphone. Technological proficiency is therefore not just a personal trait but often a function of socioeconomic status, educational background, and public infrastructure.

- **Digital Divide**: In many parts of the Global South, broadband internet and mobile networks are not always robust. This directly restricts access to online health apps or wearable monitoring devices. In rural India, for instance, only a small fraction of older adults own smartphones. This discrepancy means that an advanced telehealth solution might not be relevant unless adapted to low-tech or offline environments.
- **Language Barriers**: Many AgeTech interfaces default to English, which can alienate older adults who speak regional languages. Even within a single country, there may be

multiple languages and dialects. Unless designers include multilingual options or simple icons, older adults unfamiliar with the dominant languages may find the technology inaccessible or intimidating.

18.2.3 Societal Attitudes Toward Technology

Across my travels, I have observed that communities differ in their underlying attitudes toward new technologies. Some groups embrace innovation enthusiastically, while others exercise caution due to prior negative experiences or broader concerns over privacy.

- **Technophilia versus Technophobia**: In societies that celebrate high-tech progress, older adults might receive training and encouragement from younger generations, fostering a culture of experimentation with AgeTech. Alternatively, in places where skepticism towards new devices runs deep, there may be immediate suspicion that any form of monitoring or data collection infringes on personal boundaries.
- **Trust in Technology**: Trust can be influenced by historical or political contexts. Regions with a history of surveillance or data misuse might be wary of new digital tools. AgeTech solutions, particularly those involving constant health monitoring, could heighten fears about personal privacy and misuse of data, slowing adoption.

18.3 Case Studies from the Global South

18.3.1 AgeTech in China: Balancing Tradition and Modernity

China's aging population has grown rapidly due in part to the one-child policy. While traditional values of filial piety remain strong, urbanization has left many older adults living apart from their children [9]. Some families turn to smart home systems that track daily activities and vital signs, offering peace of mind to distant relatives. However, acceptance is uneven. I recall speaking with a Chinese colleague who mentioned that her grandparents felt uneasy relying too heavily on "machines" instead of family members. This resonates with studies showing that while some older adults appreciate the convenience of smart homes, others feel that technology can never replicate the warmth of human care.

Robotic caregivers such as Xiao Bao are also emerging in nursing homes, assisting with basic tasks and providing companionship. Yet the idea that a robot might replace close personal contact clashes with the deep cultural emphasis on family. The general consensus is that these robots are viewed more positively if framed as supplements to, rather than substitutes for, human interaction.

18.3.2 AgeTech in India: Addressing the Digital Divide

India's demographic diversity creates a patchwork of opportunities and barriers for AgeTech. In metro cities like Mumbai or Bangalore, smartphone usage is common among older adults, but rural areas tell a different story. During a research visit to a rural district, I noticed that even though older adults might own feature phones, they rarely accessed health apps because of low literacy and language issues.

- **Mobile Health Apps**: Applications like "mMitra," originally designed for maternal and child health, have been adapted to serve older adult health needs as well. These apps use local languages and audio messages to overcome literacy gaps.
- **Community-Based Interventions**: The "Elder Spring Response System" exemplifies how simple mobile networks can connect older adults in remote areas to community help. Here, technology acts as a bridge rather than a standalone solution, reinforcing local care networks rather than displacing them.

18.3.3 AgeTech in Latin America: Social Inclusion and Accessibility

Countries like Brazil and Mexico have made significant strides in AgeTech to promote social inclusion and healthcare access. In Brazil, government-funded digital literacy programs help older adults gain confidence online, resulting in a tangible sense of empowerment. I find these programs especially inspiring because they emphasize collective learning; for example, local youth sometimes volunteer to teach older adults digital basics, strengthening intergenerational bonds.

Telemedicine also plays a crucial role. In Mexico, remote healthcare services target older adults in rural regions. These efforts highlight the importance of culturally sensitive interface design—for instance, including icons that resonate with local practices or providing instructions in indigenous languages alongside Spanish.

18.3.4 AgeTech in Africa: Innovative Solutions Amidst Challenges

African nations face common challenges such as infrastructural limitations and lower digital literacy. However, I have been consistently impressed by the resourcefulness and innovation emerging from the continent's tech sector.

- **Mobile Banking for the Older Person**: In Kenya, M-Pesa has transformed financial transactions. Older adults can manage pensions or send money to family without the need for traditional banking [10]. As one older adult described it to me, using M-Pesa is far simpler than traveling to the nearest bank branch, which could be hours away.

- **Assistive Technologies**: In South Africa, low-cost assistive devices have been developed with input from local communities. This bottom-up approach ensures that the final product reflects real needs, such as navigating uneven outdoor terrain or living in multi-generational households.

18.4 Designing Culturally Appropriate AgeTech Solutions

18.4.1 User-Centered Design with Cultural Sensitivity

One lesson that recurs in my practice is that older adults are experts in their own lived experiences. When designers partner with them from the outset, technologies are far more likely to align with local customs and day-to-day realities [8]. Additionally, appreciating symbolic colors, communal norms, and religious taboos can avert the inadvertent alienation of certain groups.

18.4.2 Overcoming Language and Literacy Barriers

- **Multilingual Interfaces**: Offering local language interfaces immediately widens a technology's reach. In my fieldwork, I have seen how voice-activated systems can transform usability for older adults who are not proficient readers.
- **Simplified Designs**: Reducing on-screen text and relying on intuitive icons helps seniors who have limited formal education. Research also confirms that universal symbols, large buttons, and simplified navigation aid older users significantly.

18.4.3 Building Trust and Addressing Privacy Concerns

Promoting transparency about data collection is vital. Older adults often worry about who can see their personal information or health records. Demonstrating that data is secure and used responsibly can quell fears [1]. Community leaders, whether religious figures, local politicians, or respected elders, can also help advocate for a given technology, leveraging existing networks of trust [6].

18.4.4 Economic Considerations

- **Affordability**: Many older adults live on modest or fixed incomes. Low-cost or shared devices can ease financial barriers. I once witnessed a co-op in which multiple families pooled resources to buy a single assistive device that everyone could access when needed.

- **Sustainable Models**: Partnerships with governments or NGOs can ensure solutions are not abandoned due to high maintenance costs. The World Health Organization (2015) notes that successful public–private collaborations can safeguard the long-term viability of AgeTech initiatives, especially in underserved communities.

18.5 Impact of Cultural Differences on Specific AgeTech Domains

18.5.1 Telehealth and Telemedicine

In some cultures, a face-to-face conversation with a doctor is considered integral to quality care, leading to skepticism about telehealth. When I shadowed a telemedicine project in East Africa, older adults expressed concern that remote doctors would not truly understand their conditions unless they met physically. Cultural adaptations, such as combining video visits with in-person checkups or involving local community health workers, have shown promise in overcoming this resistance [12].

18.5.2 Robotics and AI in Older Person Care

Robots often evoke strong emotional responses. While Japan embraces social robots, other regions may perceive them as impersonal or even eerie [11]. One practical solution is to customize a robot's appearance and mannerisms to reflect local cultural norms—perhaps adopting a gentle tone, using culturally appropriate greetings, or wearing attire that aligns with local customs [5].

18.5.3 Wearable Technologies

Wearables like fitness trackers can raise privacy and modesty concerns. During a visit to a conservative region, I noticed older adults were reluctant to wear a device on their wrist that publicly displayed health data [7]. Designing unobtrusive devices that fit discreetly under clothing, or using culturally neutral aesthetics, can make these products more acceptable [9].

18.6 Strategies for Successful Implementation of AgeTech in Diverse Cultural Contexts

18.6.1 Education and Training

Digital literacy programs that teach seniors to navigate smartphones or specialized devices can significantly reduce resistance. I once volunteered at a workshop where teenagers taught older adults how to text and browse the internet, and the enthusiasm and connection sparked between generations was remarkable. In addition, intergenerational learning, where younger relatives assist their elders, can build family unity and break down technological barriers, further easing the adoption of new tools [2].

18.6.2 Policy and Government Support

Governments have a pivotal role in making AgeTech accessible. Inclusive policies that recognize older adults' unique needs, coupled with subsidies or tax breaks for AgeTech products, can elevate adoption rates [13]. Adequate funding and resource allocation, particularly for rural and low-income areas, further ensures that AgeTech does not become a luxury reserved for wealthier populations [14].

18.6.3 Collaboration with Local Organizations

Partnering with grassroots organizations helps tailor solutions to local values and constraints [3]. Nonprofits, community groups, and religious institutions often have deep trust within their communities. Co-design methods that engage local leaders from the beginning of a project can yield AgeTech tools more aligned with real-world circumstances.

18.7 Conclusion

My journey exploring AgeTech solutions across various cultural landscapes has repeatedly shown me that technology's power lies not just in its innovations but in its capacity to align with local traditions and mindsets. Cultural differences profoundly shape how older adults perceive technology, whether they trust it, and ultimately whether they will use it. By examining case studies from the Global South, such as China's balancing of filial piety with urban realities, India's efforts to bridge the digital divide, Latin America's emphasis on social inclusion, and Africa's resourceful solutions, we see how essential it is to weave cultural insights into design.

Developing culturally appropriate AgeTech means engaging older adults themselves, involving local communities in design decisions, and respecting language, family structures, and social norms. From multilingual interfaces to intergenerational learning and government-backed support, there are many pathways to ensure technology genuinely improves the lives of seniors. I have personally observed how empathy, community collaboration, and sensitivity to cultural nuances can transform a technological prototype into a meaningful solution that resonates with the very people it aims to help. Ultimately, it is by embracing these diverse cultural factors that AgeTech can fulfill its promise of enhancing the well-being and dignity of older adults around the world.

References

1. Bennett, C. J. (2011). In defense of privacy: The concept and the regime. *Surveillance & Society, 8*(4), 485–496. https://doi.org/10.24908/ss.v8i4.4184
2. Chen, Y.-R. R., & Schulz, P. J. (2016). The Effect of Information Communication Technology Interventions on Reducing Social Isolation in the Elderly: A Systematic Review. *J Med Internet Res, 18*(1), e18. https://doi.org/10.2196/jmir.4596
3. Chib, A. (2013). The promise and peril of mHealth in developing countries. *Mobile Media & Communication, 1*(1), 69-75. https://doi.org/10.1177/2050157912459502
4. Choudrie, J., Ghinea, G., & Songonuga, V. N. (2013). Silver Surfers, E-government and the Digital Divide: An Exploratory Study of UK Local Authority Websites and Older Citizens. *Interacting with Computers, 25*(6), 417-442. https://doi.org/10.1093/iwc/iws020
5. de Graaf, M. M. A., & Ben Allouch, S. (2013). Exploring influencing variables for the acceptance of social robots. *Robotics and Autonomous Systems, 61*(12), 1476-1486. https://doi.org/10.1016/j.robot.2013.07.007
6. Kamal, S. A., Shafiq, M., & Kakria, P. (2020). Investigating acceptance of telemedicine services through an extended technology acceptance model (TAM). *Technology in Society, 60*, 101212. https://doi.org/10.1016/j.techsoc.2019.101212
7. Lee, C., & Coughlin, J. F. (2015). PERSPECTIVE: Older Adults' Adoption of Technology: An Integrated Approach to Identifying Determinants and Barriers. *Journal of Product Innovation Management, 32*(5), 747-759. https://doi.org/10.1111/jpim.12176
8. LeRouge, C., Ma, J., Sneha, S., & Tolle, K. (2013). User profiles and personas in the design and development of consumer health technologies. *International Journal of Medical Informatics, 82*(11), e251-e268. https://doi.org/10.1016/j.ijmedinf.2011.03.006
9. Luo, B., & Zhan, H. (2012). Filial Piety and Functional Support: Understanding Intergenerational Solidarity Among Families with Migrated Children in Rural China. *Ageing International, 37*(1), 69-92. https://doi.org/10.1007/s12126-011-9132-1
10. Ngugi, B., Pelowski, M., & Ogembo, J. G. (2010). M-pesa: A Case Study of the Critical Early Adopters' Role in the Rapid Adoption of Mobile Money Banking in Kenya. *THE ELECTRONIC JOURNAL OF INFORMATION SYSTEMS IN DEVELOPING COUNTRIES, 43*(1), 1-16. https://doi.org/10.1002/j.1681-4835.2010.tb00307.x
11. Papadopoulos, I., Koulouglioti, C., Lazzarino, R., & Ali, S. (2020). Enablers and barriers to the implementation of socially assistive humanoid robots in health and social care: a systematic review. *BMJ Open, 10*(1), e033096. https://doi.org/10.1136/bmjopen-2019-033096

12. Shiferaw, K. B., & Mehari, E. A. (2019). Modeling predictors of acceptance and use of electronic medical record system in a resource limited setting: Using modified UTAUT model. *Informatics in Medicine Unlocked, 17*, 100182. https://doi.org/10.1016/j.imu.2019.100182
13. United Nations. (2024). World Population Prospects 2024. Retrieved from https://www.un.org/development/desa/pd/world-population-prospects-2024
14. WHO. (2015). *World report on aging and health.* Retrieved from Geneva: https://www.who.int/publications/i/item/9789241565042/

Education and Training for the Next Generation of AgeTech Professionals

19

19.1 Introduction

AgeTech spans diverse innovations, including assistive devices, health-monitoring tools, and social engagement platforms [9]. Whether it is a wearable device that records vital signs or an application designed to foster intergenerational connections, AgeTech is poised to transform how older adults live and interact with their communities. Yet, as I have learned through my involvement in technology-focused projects, building gadgets is only part of the solution. What we really need is a dedicated workforce that not only possesses the requisite technical acumen but also understands older adults' unique needs, preferences, and cultural contexts. That realization underpins why specialized education and training in AgeTech has become so essential.

In this essay, I explore how we can best equip future professionals for the AgeTech sector. We will begin by looking at demographic changes and how they shape the need for specialized education. From there, we'll delve into the various models of learning that can help students master both theoretical and practical competencies. Programs like Early Professionals, Inspired Careers (EPIC), under the AGE-WELL network, offer us concrete examples of how these strategies come together. Ultimately, my goal is to highlight the comprehensive approaches, combining interdisciplinary coursework, experiential learning, and continuous professional development, that are necessary to build a proficient and empathetic AgeTech workforce.

A. Mihailidis, *The Age of AgeTech*, Synthesis Lectures on Technology and Health,
https://doi.org/10.1007/978-3-032-04886-8_19

19.2 The Need for Specialized Education in AgeTech

19.2.1 Demographic Changes and Aging Population

As the World Health Organization [13] points out, the rapidly growing older adult population implies new social and economic realities that will shape public policy, healthcare systems, and housing markets, among other domains. When I reflect on how my own grandparents prefer to age in place, I see how AgeTech can be the linchpin, helping them stay connected, safe, and healthy. Yet deploying technological solutions on such a massive scale demands professionals who can navigate both advanced digital systems and the nuances of gerontology. Indeed, the success of AgeTech in improving older adults' quality of life will hinge on whether we have enough individuals with deep, interdisciplinary expertise.

Given these demographic changes, specialized training in AgeTech goes beyond merely teaching students how to write code or design circuit boards. It involves understanding physical changes that occur with aging, like reduced mobility or sensory impairments, as well as cognitive shifts that might influence how older adults engage with technology. Equally important are social dynamics—how older adults might experience isolation, or how they often rely on family networks for decision-making. Understanding these aspects helps professionals tailor technologies that genuinely resonate with users.

19.2.2 Technological Advancements and Opportunities

At the same time, emerging technologies, ranging from AI-driven applications to the Internet of Things (IoT), are transforming what is possible in AgeTech. I have been particularly intrigued by how wearables can monitor health metrics in real time, alerting caregivers if an older adult's vital signs dip. Meanwhile, robotics and telehealth platforms reduce the burden of travel for those who find it challenging to leave home. However, such tools are only as impactful as the professionals behind them. Without well-trained experts, we risk developing technologies that look great on paper but fail in practice because they do not account for older adults' lived realities.

19.2.3 Skill Requirements in AgeTech

What, then, does a competent AgeTech professional look like? First and foremost, they need to understand user-centered design so that the tools they create or deploy are accessible and intuitive. They also need a grasp of gerontological principles: how diseases like dementia progress, or how certain interventions can improve mental well-being. Policy awareness is another must. For instance, if a device collects personal health data, the professional should be conversant in privacy regulations and ethical guidelines [7]. Finally, empathy is indispensable. From personal experience, I've seen how older adults often feel vulnerable when adopting new technology, especially if they have had minimal prior exposure. A professional who can empathize and patiently guide them through these devices can make all the difference in boosting adoption and satisfaction.

19.3 Models of Education and Training in AgeTech

19.3.1 Theoretical Learning

19.3.1.1 University Courses and Degree Programs

Academic institutions play a critical role in laying the groundwork. Students who enroll in gerontology, computer science, engineering, or design programs gain the theoretical foundations needed to innovate responsibly. It is not uncommon to find individual courses scattered across different departments, say, a computer science course focusing on machine learning for healthcare, or a gerontology course that touches upon technological aids. Some universities are going a step further, integrating these components into specialized programs that merge technology and aging studies.

For example, the University of Southern California offers a Master's program in Gerontology with a strong emphasis on technology's intersection with aging (USCLeonard Davis). Students in such programs not only absorb the fundamentals of gerontology, learning about age-related changes and societal implications, but also explore specific technological applications, such as assistive robotics or digital health platforms. From my perspective, these hybrid courses are vital in showing students the "big picture," ensuring they do not just excel at coding but also comprehend the broader context in which their solutions will be used.

19.3.1.2 Online Courses and MOOCs

Beyond brick-and-mortar institutions, Massive Open Online Courses (MOOCs) have democratized access to AgeTech education [11]. During the COVID-19 pandemic, I personally enrolled in an online course that focused on user interface design for older adults. Despite the remote setting, the curriculum was interactive, featuring case studies and discussion forums where learners from around the globe shared their experiences. By uniting participants across different countries and professional backgrounds, MOOCs create a rich, cross-pollinating environment. This is especially advantageous for people working full time or living in rural areas who might not be able to attend in-person classes.

19.3.2 Experiential Learning

19.3.2.1 Internships and Co-Op Programs

While theoretical knowledge lays the foundation, hands-on experience often cements those lessons into practical skill sets. Internships or co-op opportunities with AgeTech startups, research labs, or healthcare institutions allow students to face real-world scenarios they cannot replicate in a standard classroom [4]. As an intern in a wearable device startup, I witnessed how critical it is to test prototypes with actual older adult users, who might point out usability issues the design team never anticipated. Such firsthand learning fosters adaptability and sharpens problem-solving skills, traits that theoretical study alone rarely imparts.

19.3.2.2 Practical Projects and Lab Work

University labs and innovation hubs dedicated to Ambient Assisted Living or other AgeTech-focused research provide fertile ground for experiential learning. Students here can work in interdisciplinary teams, combining software engineering, industrial design, and clinical insights to tackle real-life challenges, like developing a smart sensor that detects falls or a VR application that aids physical therapy. From my experience, these team-based projects are invaluable for honing communication skills; engineers learn to speak the language of occupational therapists, while design students discover the medical constraints around certain conditions. The result is a more cohesive, human-centered approach.

19.3.2.3 Mentorship and Networking Opportunities

Mentorship can be a pivotal factor in shaping one's career trajectory. I have been fortunate to find mentors in both academic and industry settings who helped me understand not just the technical challenges but also how to navigate the complexities of user adoption and funding. Networking events, conferences, and professional associations similarly open doors, allowing budding AgeTech professionals to stay current on trends and forge partnerships [2]. In AgeTech, where research and commercial interests often overlap, these connections can lead to collaborations that accelerate innovation.

19.4 Examples of Training Programs in AgeTech

19.4.1 EPIC in AGE-WELL

AGE-WELL is Canada's technology and aging network dedicated to fostering innovations that improve the lives of older adults and caregivers (AGE-WELL). Funded by the Networks of Centres of Excellence program, it brings together academia, industry, and community stakeholders to drive research and commercialization. Through my involvement with AGE-WELL events, I have observed how the network actively engages older adults themselves, treating them as essential co-creators rather than passive recipients of technology.

One standout initiative within AGE-WELL is Early Professionals, Inspired Careers (EPIC), a training program that helps cultivate the next generation of AgeTech leaders (AGE-WELL). EPIC's framework includes.

- **EPIC Core Program**: Foundational training in interdisciplinary research methods, innovation strategies, and knowledge mobilization. This ensures that participants develop not only technical skills but also the capacity to translate research findings into tangible products or policies.
- **Workshops and Webinars**: Covering topics as diverse as commercialization tactics, policy considerations, and ethical standards. I recall attending one such webinar on data privacy in telehealth, which drove home how essential transparency and informed consent are when rolling out digital health tools.
- **Internships and Entrepreneurial Opportunities**: EPIC trainees often gain hands-on exposure by working directly with industry or community partners. This synergy between academia and real-world practice deepens their skill sets and fosters new relationships.
- **Mentorship and Networking**: The program connects students with seasoned professionals. In my view, this is a critical component because it fosters knowledge exchange and encourages aspiring AgeTech professionals to learn from veterans in the field.

Since its inception, EPIC has trained hundreds of participants, many of whom have branched out into academia, industry, or their own startups [12]. From my conversations with some of these graduates, it is clear that the program's emphasis on holistic skill-building, technical, ethical, and interpersonal, equips them to tackle challenges creatively. Whether they end up pioneering new smart home technology or improving data analytics in long-term care settings, these individuals carry forward a blend of research rigor and practical know-how.

19.4.2 Other Notable Programs

19.4.2.1 Gerontechnology Master's Programs

Similar efforts are emerging outside North America. The University of Maastricht in the Netherlands, for instance, offers a Master's in Gerontology with a strong focus on technological applications. These specialized programs underscore a growing consensus that the study of aging and technology must be interwoven, not treated as separate spheres.

19.4.2.2 Specialized Courses in Universities

Single courses like "Technology and Aging" are appearing in engineering and health science departments around the world [10]. Although these courses may vary in scope, they generally address fundamental design principles for older users, usability testing, and the physiological aspects of aging. In my experience, even one well-structured course can ignite students' passion for AgeTech, prompting them to seek internships or research opportunities in the field.

19.4.2.3 Industry-Led Training Initiatives

Companies specializing in AgeTech sometimes offer internal training for new employees or collaborate with educational institutions on targeted curricula. This approach ensures that education aligns with industry demands, fostering a workforce that can hit the ground running. I have seen job listings requiring AgeTech experience, confirming that such specialized knowledge is increasingly valued. Industry-driven programs also serve as a reality check for academic theories, ensuring that they align with market and user needs.

19.5 Strategies for Preparing Future AgeTech Professionals

19.5.1 Interdisciplinary Education

One of the central lessons I have gleaned from my foray into AgeTech is that no single discipline holds all the answers. Technologies for older adults require input from gerontology, engineering, design, sociology, and more [8]. Interdisciplinary programs cultivate an environment where students learn to see a problem from multiple angles. For instance, an engineering student might collaborate with a public health student, and in the process, they discover the importance of policy and community engagement.

Case Study: Interdisciplinary Curriculum Development—At the University of Toronto, a course invites engineering and health science students to form teams tasked with creating AgeTech prototypes. Not only do these teams merge distinct skill sets, but they also learn to reconcile different vocabularies and methodological approaches. I have met a few alumni from this course, and they all cite the collaborative environment as key to their professional development. They mention it taught them how to articulate technical constraints to non-engineers while understanding the nuanced health challenges that older adults face.

19.5.2 Collaboration Between Academia and Industry

Tight-knit partnerships between universities and industry can significantly elevate AgeTech training. Companies stand to benefit from fresh ideas and research-based insights, while students gain practical exposure to real-world constraints and market dynamics. Personally, I recall a co-op project that placed me in an AgeTech firm specializing in fall detection devices. The experience shed light on how iterative product development, based on user feedback and robust testing, directly influences the final product's reliability and acceptance. Such opportunities equip students with a level of professional polish that is hard to replicate in traditional classroom settings.

Example: Co-op Programs with AgeTech Companies—When a student cycles between academic terms and work placements at an AgeTech startup, they can see how theoretical principles, like software architecture or sensor calibration, translate into practical solutions. The student also returns to campus with real project data, fueling deeper academic projects or thesis work. It is a virtuous cycle that benefits educational institutions, employers, and, ultimately, older adults who receive better designed products and services.

19.5.3 Incorporating User-Centered Design and Ethics

Technological solutions for older adults can fail spectacularly if they don't align with the users' actual needs and preferences. Throughout my own projects, I have witnessed how easy it is to overlook simple barriers, like small text on an app screen or confusing menu navigation. To counter this, courses that emphasize user-centered design principles and ethics can shape more responsible innovators.

19.5.4 Ethical Considerations

Many older adults are sensitive about data privacy, consent, and the possibility of being monitored continuously. Educational programs addressing these ethical aspects instill a mindset where students are taught to balance innovation with respect for users' autonomy [1]. Beyond just checking legal boxes, they learn to engage older adults in meaningful dialogues about how data is gathered and used.

19.5.5 User-Centered Design Projects

Integrating projects where students interact directly with older adults helps them cultivate empathy. For example, they might do usability testing at a local senior center, collecting feedback on a prototype. Such experiences underscore the importance of iterative design and remind students that technology must serve, not dictate, the users' lives [6].

19.6 Challenges and Solutions

19.6.1 Keeping Curriculum up to Date with Rapid Tech Changes

Given how quickly AI, robotics, and data analytics evolve, AgeTech courses can become outdated if universities rely on static curricula. In my own academic experiences, I have seen courses linger on technologies that are no longer industry-standard, leaving graduates less prepared for current demands.

Solution: Continuous Curriculum Review—Regular curriculum revisions informed by an advisory board of industry professionals can keep educational content fresh. I have also seen programs that incorporate annual "tech showcases" where local businesses present emerging tools, thereby exposing students and faculty to the latest breakthroughs.

19.6.2 Ensuring Accessibility and Inclusivity in Education

Cost, geographic location, and educational prerequisites can deter some potential learners from AgeTech programs. This is especially crucial in areas where interest in AgeTech is high, but local institutions lack robust infrastructure.

Solution: Online Learning Platforms—Virtual courses break down many of these barriers [5]. As someone who has benefited from remote learning, I can attest that a well-structured online platform can be just as engaging and comprehensive as on-campus alternatives. Incorporating live Q&A sessions, discussion forums, and group projects further ensures an interactive experience.

19.6.3 Overcoming Barriers to Experiential Learning

Physical internships or lab placements may be limited by geographic constraints, funding shortages, or organizational capacity.

Solution: Virtual Simulations and Projects—Universities can adopt simulated environments, like VR labs or advanced software, that replicate real-world conditions [3]. For instance, a virtual scenario might let students experiment with sensor networks in a "smart home" simulation, tracking an avatar's movements to detect potential design pitfalls. While not a perfect substitute for in-person experiences, these simulations broaden training access substantially.

19.7 Conclusion

Preparing the next generation of AgeTech professionals has never felt more urgent. Every conversation I have with researchers, healthcare providers, or older adults themselves reaffirms the need for technology that genuinely enhances day-to-day life for seniors. Yet building that technology, and ensuring it is adopted widely, requires a workforce skilled in the complexities of both aging and innovation. Through interdisciplinary coursework, immersive internships, mentorship, and continuous curriculum updates, we can forge a pipeline of talent equipped to address the demands of an aging global population.

Initiatives like the Early Professionals, Inspired Careers (EPIC) program within AGE-WELL exemplify how academia, industry, and community stakeholders can unite to nurture these skills. By blending theoretical foundations in gerontology and emerging tech with hands-on experiences, EPIC has demonstrated meaningful results in shaping driven, empathetic individuals who go on to succeed in academia, entrepreneurship, or corporate roles in the AgeTech arena. Similar approaches—from the Master's programs in Gerontechnology to industry-led short courses—reinforce the collective push toward a future where older adults benefit from technology that is thoughtfully designed and ethically implemented.

Looking ahead, collaboration will be key. Educational institutions must work in tandem with AgeTech companies to align curricula with real-world opportunities and challenges. Policymakers can encourage these partnerships, ensuring that future professionals have the resources, incentives, and frameworks needed to excel. And in a field where technology evolves so rapidly, lifelong learning stands out as an essential mindset. Ultimately, by investing in these robust educational pathways, we not only equip professionals to keep pace with technological shifts but also fulfill our responsibility to enhance the well-being of a generation that deserves dignified, supportive, and empowering solutions.

References

1. Brey, P. A. E. (2012). Anticipatory Ethics for Emerging Technologies. *NanoEthics, 6*(1), 1-13. https://doi.org/10.1007/s11569-012-0141-7
2. Johnson, W. B. (2015). *On being a mentor: A guide for higher education faculty*: Routledge.
3. De Jong, T., Linn, M. C., & Zacharia, Z. C. (2013). Physical and Virtual Laboratories in Science and Engineering Education. *Science, 340*(6130), 305-308. https://doi.org/10.1126/science.1230579
4. Kolb, D. A. (2014). *Experiential learning: Experience as the source of learning and development*: FT press.
5. Means, B., Bakia, M., & Murphy, R. (2014). *Learning online: What research tells us about whether, when and how*: Routledge.
6. LeRouge, C., & Wickramasinghe, N. (2013). A Review of User-Centered Design for Diabetes-Related Consumer Health Informatics Technologies. *Journal of Diabetes Science and Technology, 7*(4), 1039-1056. https://doi.org/10.1177/193229681300700429

7. Neves, B. B., & Amaro, F. (2012). Too old for technology? How the elderly of Lisbon use and perceive ICT. *The journal of community informatics, 8*(1), 1-12.
8. Newell, W. H. (2013). The State of the Field: Interdisciplinary Theory. *Issues In interdisciplinary studies, 31*, 22-43.
9. Peek, S. T. M., Wouters, E. J. M., van Hoof, J., Luijkx, K. G., Boeije, H. R., & Vrijhoef, H. J. M. (2014). Factors influencing acceptance of technology for aging in place: A systematic review. *International Journal of Medical Informatics, 83*(4), 235-248. https://doi.org/10.1016/j.ijmedinf.2014.01.004
10. Rogers, W. A., & Fisk, A. D. (2010). Toward a Psychological Science of Advanced Technology Design for Older Adults. *The Journals of Gerontology: Series B, 65B*(6), 645-653. https://doi.org/10.1093/geronb/gbq065
11. Zhang, K., Bonk, C., Reeves, T., & Reynolds, T. (2019). *MOOCs and Open Education in the Global South: Challenges, Successes, and Opportunities*: Taylor & Francis.
12. AGE-WELL. (2020). Annual Report 2020. Retrieved from https://agewell-nce.ca/publications/annual-reports
13. WHO. (2015). *World report on aging and health.* Retrieved from Geneva: https://www.who.int/publications/i/item/9789241565042/

20 Will We Ever See Anything Novel Again in AgeTech?

20.1 Introduction

The world's demographic fabric is unravelling before our eyes. By 2050, the number of people aged 60 and above will skyrocket to 2 billion, according to the World Health Organization [11]. This surge in older adults offers tantalizing opportunities for technological breakthroughs—yet also underscores colossal challenges that most of us in the AgeTech arena have barely started to reckon with. Sure, we have seen countless headlines about "innovations" in caregiving robots, wearables that promise to track health better, and telehealth systems that vow to revolutionize medical consultations. But if we are brutally honest, how many of these solutions are genuinely groundbreaking?

I have spent enough time in AgeTech conferences and research labs to note a sobering reality: we are drowning in incremental tweaks rather than experiencing seismic shifts. It is like we are all patting ourselves on the back for making the same solutions slightly shinier and, in the process, ignoring the radical leaps that could genuinely redefine aging. This essay aims to drag these issues into the spotlight and pose a blunt question: Will we ever see anything novel again in AgeTech?

I am aware that such a challenge might ruffle feathers, including my own. After all, there's a tendency to celebrate every baby step forward, every new sensor or updated app feature, as a major accomplishment. But at what cost? By over-investing in user-centred design (to the point of ignoring our professional insights) and clinging to established frameworks, we might be smothering the very disruptive ideas that could achieve true transformation. This essay is a call to break free from the safe path of incrementalism and to trust our expertise enough to chase after the bold visions that older adults genuinely need.

A. Mihailidis, *The Age of AgeTech*, Synthesis Lectures on Technology and Health,
https://doi.org/10.1007/978-3-032-04886-8_20

20.2 The State of AgeTech: An Era of Incrementalism

20.2.1 A Proliferation of Redundancy

There was a time, roughly two decades ago, when AgeTech felt like a frontier of unlimited promise. Innovations such as smart home sensors, wearable devices, and telehealth platforms seemed poised to alter the entire aging experience. Yet fast-forward to today, and what have we got?

A dizzying array of redundant offerings. Consider the universe of fall detection devices. Let's be real: while fall prevention is undeniably crucial, the differences between these devices mostly revolve around aesthetics, minor improvements in sensor accuracy, or app interface tweaks [3]. Rarely do we encounter a system that radically challenges the underlying assumptions of how we approach fall prevention. Instead, we see the same old problem addressed with nearly identical solutions—each marketed as the "next big thing."

I cannot help but feel frustrated when I walk through an AgeTech expo, scanning booth after booth of technologies that promise "revolution" but deliver only incremental refinements. It is not that small improvements do not matter; they do. However, our collective willingness to settle for these small wins signals a kind of complacency. It begs a deeper question: Are we losing sight of the bigger picture in the quest for short-term gains?

20.2.2 The Comfort of the Known

Part of this incrementalism stems from a deep-seated preference for safety. It is easier to refine existing devices than to attempt something truly radical, especially when the latter carries risks that might scare off investors or regulators [2]. For instance, if you are pitching a wearable that is just slightly better than the one that came before, you can rely on established user acceptance and a smoother path through regulatory bodies.

But let's acknowledge that this comfort zone can stifle the creative leaps that we so badly need. I see it often: a startup gets a whiff of success with one product, and suddenly every iteration thereafter revolves around that single concept, incrementally optimized rather than boldly reimagined. Yes, it is a sensible business strategy in many fields, but in AgeTech, where human lives and well-being hang in the balance, shouldn't we be aiming higher than "safe bets"?

20.3 The Innovator's Dilemma in AgeTech

20.3.1 AgeTech's Entrapment in Sustaining Innovations

Clayton Christensen's *The Innovator's Dilemma* (1997) argued that established organizations often fail to embrace disruptive innovations because they are too busy perfecting existing products for current customers. AgeTech appears to be a textbook case of this phenomenon. We talk endlessly about co-creation with older adults—surveying them, holding focus groups, gathering user feedback at every turn [8]. On the surface, this approach seems noble and necessary. Yet in practice, it can anchor us to known variables.

Let's not forget that older adults, like all users, often draw upon their existing experiences to frame their feedback. If a certain style of device or interaction is all they know, they may resist something that demands a radically different approach. It is like trying to introduce a high-end virtual reality platform to individuals who are comfortable with phone calls and FaceTime [10]. The immediate response might be lukewarm or even negative, not necessarily because the tech is flawed, but because it demands a shift in mindset.

So, while user-centered design has a rightful place, relying on it exclusively can blind us to the potential for radical, disruptive innovation. Ironically, our zeal to "listen to the users" can morph into a cage of limited ambitions, where we make incremental adjustments rather than radical changes.

20.3.2 The Risk of Disruption from Outside

To me, one of the most glaring dangers of AgeTech's sustaining-innovation trap is the possibility of being blindsided by outsiders who could not care less about traditional assumptions. Picture tech giants swooping into the space with unlimited R&D budgets and no emotional investment in the old ways. Or perhaps it's a maverick startup, unburdened by conventional gerontological thinking, that unleashes a solution so different and compelling that existing AgeTech players look prehistoric by comparison.

We have seen this story unfold in other industries: think of how Tesla took on the automotive giants or how Netflix toppled Blockbuster. AgeTech is not immune to such a revolution. Indeed, given the appetite for solutions that handle the complexities of aging, it might just be a matter of time before a seismic disruption knocks everyone else off the board.

20.4 The Need for Novelty: Beyond Incremental Improvements

20.4.1 Reimagining Problems and Solutions

If there is one thing I have learned from working in AgeTech, it is that the problems we are tackling are often more complex than they appear. A medication reminder app is useful, but it does not fundamentally address why older adults might miss doses, maybe it's loneliness, maybe it's memory issues, or perhaps they distrust medical advice after a bad experience. At some point, another "improved medication reminder" starts to look like superficial tinkering.

What if, instead, we tackled the root causes? Imagine bio-integrated sensors that continuously analyze bodily markers and autonomously adjust medication dosages—way beyond a beep on the phone [5]. Yes, that is daunting, especially from a regulatory and ethical standpoint. But it is also the kind of leap that could upend how we think about aging and health management.

This shift requires us to question our basic assumptions and to stop seeing older adults only as end-users who need help with daily tasks. Maybe they want to remain active contributors to society. Maybe they have opinions or tastes that we have not explored yet, like how older adults might want wearable devices that are stylish or that reflect cultural values. We do ourselves a disservice when we only tweak old ideas. We should be overhauling the entire concept of how tech intersects with aging.

20.4.2 Embracing Failure as a Pathway to Innovation

If we are serious about forging breakthroughs, we must accept that failure is a companion on this journey [4]. I have encountered funders and institutions who balk at any whiff of risk. They want guaranteed outcomes, metrics, and user acceptance. But breakthroughs rarely emerge from sanitized projects. They come from bold leaps into the unknown, where half-baked prototypes might flop spectacularly before they morph into something truly groundbreaking.

Consider historical examples: the first personal computers were not exactly crowd-pleasers, and early smartphones were ridiculed by many. Without a willingness to fail, we would not have the ubiquitous tech that shapes our modern world. In AgeTech, a "safe" culture that discourages risk effectively guarantees a pipeline of incremental products. Perhaps we need to invert our success metrics to reward visionary attempts, even if they occasionally result in dead ends.

20.5 Trusting the Experts: Valuing Professional Insight

20.5.1 The Limits of User-Centered Design

I realize it is almost heretical in modern design thinking to critique user-centred methods, but let's be frank: users do not always know what they want, especially when the technology in question may be unlike anything they have experienced [9]. Older adults provide invaluable perspectives on usability and relevance, but they might have no frame of reference for solutions that go well beyond their immediate experiences.

In AgeTech, this dynamic is particularly poignant because older adults can be more cautious with new technology, sometimes even skeptical. Constantly pivoting to satisfy current user preferences can hamper the potential for truly transformative concepts. It is somewhat analogous to Henry Ford's purported remark: "If I had asked people what they wanted, they would have said faster horses."

20.5.2 The Role of Visionary Leadership

We love touting the importance of collaboration and user input, but what about visionary leaders who champion ideas even when users and stakeholders are initially lukewarm? Look at Steve Jobs' often-quoted take on consumer demand: "People don't know what they want until you show it to them" [6]. That statement may sound brash, but perhaps AgeTech needs a bit of that swagger.

Sometimes, professionals, engineers, gerontologists, product managers, can see beyond immediate user requests. They can anticipate needs the public cannot articulate. We have the training and domain expertise to dream bigger. If we relegate ourselves to only implementing incremental requests from end-users, we're abdicating the responsibility that comes with our specialized knowledge.

20.6 Provoking Change: A Call to Action for the AgeTech Field

20.6.1 Fostering an Innovation Ecosystem

Bringing truly novel solutions to life demands more than isolated research labs working in silos. It calls for an ecosystem where academia, industry, and government collaborate to push boundaries [7]. Instead of funneling money into the same old prototypes, public and private funders should specifically earmark grants for disruptive research—projects that might fail spectacularly but could also succeed in unimaginable ways.

Academia can champion interdisciplinary research that unites software engineers, gerontologists, social scientists, and ethicists under one roof. Why not host bold hackathons that challenge participants to rethink aging from the ground up? Why not create cross-sector consortia that break down the parochial interests of each stakeholder group?

20.6.2 Re-Evaluating Success Metrics

AgeTech's existing success metrics, such as how many units sold and how many older adults adopt the product, tend to prioritize incremental growth over radical breakthroughs [1]. If your measure of success is quick adoption, you will chase low-hanging fruit, such as slightly better apps, marginally improved sensors, because they guarantee near-term buy-in.

But what about longer-term impact? What about technologies that might upend our entire approach to caregiving or living arrangements, even if they take more time to refine? We need to shift our metrics to prioritize transformative potential, not just immediate market performance. And yes, that might mean some products won't gain traction immediately. But those that do could redefine AgeTech in ways we can barely predict now.

20.6.3 Educating the Next Generation

Ultimately, the future rests with the professionals we are training today. Having served as a mentor for budding AgeTech enthusiasts, I have seen how quickly their enthusiasm can be corralled into conventional tracks. We need curricula that invite, even demand, a radical mentality, programs that encourage risk-taking, spark interdisciplinary discussions, and situate aging in a broader socio-technical landscape.

Do not just teach students how to design a user-friendly telehealth app; push them to consider AI-driven prediction models that might supplement or even revolutionize traditional caregiving. Challenge them to think about wearable devices not just as "fitness trackers for seniors" but as gateways for older adults to remain vital in their communities, perhaps by monetizing life experiences or skills. Education should be where imagination roams free, not where it is locked into minor tweaks on existing templates.

20.7 Conclusion

The AgeTech sector currently stands at a juncture where incremental improvements have become the norm and truly revolutionary ideas appear elusive. The question, *"Will we ever see anything novel again in AgeTech?"* isn't meant to be pessimistic. Rather, it is a provocation, a challenge to all of us to scrutinize our comfortable routines and demand something more.

We have coasted on incremental changes long enough. Yes, user-centred design matters, but so does the visionary zeal to create solutions that older adults have not even dreamt of yet. Embracing failure should be recognized as an inevitable step on the road to transformative success. And trusting our professional expertise should be acknowledged as a strength, not a hindrance.

What we need now is a cultural shift, one that prioritizes big thinking and dares to disrupt. We must rally academia, industry, and policymakers behind the pursuit of bold, disruptive research. We have to realign success metrics so that radical potential isn't drowned out by the immediate gratification of user adoption. And we must groom the next generation of AgeTech innovators to be fearless, imaginative, and comfortable dancing on the edge of the unknown.

If we do this, if we truly commit to forging beyond the status quo, then the answer to our burning question is a resounding yes. We *will* see something novel again in AgeTech. We will witness breakthroughs that don't just marginally improve older adults' lives but redefine how society supports aging. It is time to look beyond incrementalism, trust our insights, and light the fires of genuine innovation in a field that needs it more desperately than ever before.

References

1. Adner, R., & Kapoor, R. (2010). Value creation in innovation ecosystems: how the structure of technological interdependence affects firm performance in new technology generations. *Strategic Management Journal, 31*(3), 306-333. https://doi.org/10.1002/smj.821
2. Christensen, C. M. (1997). *The innovator's dilemma: when new technologies cause great firms to fail.* Boston, MA: Harvard Business School Press.
3. Delahoz, Y. S., & Labrador, M. A. (2014). Survey on Fall Detection and Fall Prevention Using Wearable and External Sensors. *Sensors, 14*(10), 19806-19842.
4. Edmondson, A. C. (2011). Strategies for learning from failure. *Harvard business review, 89*(4), 48-55.
5. Heikenfeld, J., Jajack, A., Rogers, J., Gutruf, P., Tian, L., Pan, T., Wang, J. et al. (2018). Wearable sensors: modalities, challenges, and prospects. *Lab on a Chip, 18*(2), 217-248. https://doi.org/10.1039/C7LC00914C
6. Isaacson, W. (2011). *Steve Jobs*: Simon & Schuster.
7. Mazzucato, M. (2013). *The Entrepreneurial State: Debunking Public vs. Private Sector Myths*: Anthem Press.
8. Mihailidis, A., & Fernie, G. R. (2002). Context-aware assistive devices for older adults with dementia. *Gerontechnology, 2*(2), 173-188.
9. von Hippel, E. (1986). Lead Users: A Source of Novel Product Concepts. *Management Science, 32*(7), 791-805. https://doi.org/10.1287/mnsc.32.7.791

10. Waycott, J., Vetere, F., Pedell, S., Morgans, A., Ozanne, E., & Kulik, L. (2016). *Not For Me: Older Adults Choosing Not to Participate in a Social Isolation Intervention.* Paper presented at the Proceedings of the 2016 CHI Conference on Human Factors in Computing Systems, San Jose, California, USA. https://doi.org/10.1145/2858036.2858458
11. WHO. (2015). *World report on aging and health.* Retrieved from Geneva: https://www.who.int/publications/i/item/9789241565042/

Conclusion: Breaking the Cycle of Incrementalism in AgeTech

As I reflect on the essays in this collection, I am left with a clear sense of urgency and frustration. The aging population is growing faster than our innovations can keep pace, and while the AgeTech field has made progress, we seem to be trapped in a cycle of minor improvements rather than transformative breakthroughs. The time for incrementalism has passed. We need to rethink the way we approach technology for aging, or risk falling behind the very demographic we aim to serve.

AgeTech cannot afford to remain a field of small tweaks and cautious advancements. I have seen countless iterations of the same concepts, fall detection devices, medication reminders, and telehealth platforms, with little to no fundamental shift in how we think about aging. The opportunity to radically improve the lives of older adults exists, but we are too often limited by risk-averse decision-making, short-term thinking, and a reluctance to step outside of familiar territory.

This collection highlights a consistent theme: real innovation requires collaboration, but collaboration without boldness leads nowhere. AgeTech needs input from academia, industry, policymakers, and older adults themselves, but more than that, it needs courage. Aging is not a problem to solve but a reality to reshape, and that requires technology that redefines independence, health, and community engagement.

The assumption that older adults are resistant to technology is outdated. The problem is not their unwillingness to engage, but our failure to design tools that meet their needs in meaningful ways. It is time to stop designing around limitations and start designing with possibilities in mind. We need to ask tougher questions. Why are we accepting that aging inevitably leads to isolation or dependence? Why aren't we developing technologies that challenge these assumptions outright?

The essays on AI and robotics point to promising directions, but even in these fields, we seem content with incremental progress. Robotic caregivers and AI tools should not be seen as band-aids for workforce shortages but as technologies that could redefine what

A. Mihailidis, *The Age of AgeTech*, Synthesis Lectures on Technology and Health,
https://doi.org/10.1007/978-3-032-04886-8

it means to age well. The goal shouldn't just be to extend life but to improve the quality of that life in meaningful ways.

One of the most pressing issues raised in these essays is how academic and funding structures reinforce conservative thinking. I have experienced this firsthand. It is easier to secure funding for projects that refine existing solutions rather than those that propose radical departures from the norm. But innovation does not happen in safe spaces. We need to reward risk and recognize that failure is an essential part of the process. Without risk, we are left with more of the same.

I believe AgeTech needs more disruptive thinkers, entrepreneurs and researchers who are willing to break away from the status quo. We do not need another marginally better fall detection system. We need to focus on technologies that prevent falls from happening in the first place. This kind of thinking will drive real progress.

Empathy must also remain central to everything we do. AgeTech is ultimately about people. Every tool we design should begin with the question: "Would I want this for my own family?" The essays on education and workforce development reinforce this point. We need to instill empathy and user-centred design into the next generation of AgeTech professionals. Without it, we risk creating solutions that miss the mark.

I am not interested in congratulating ourselves for what has already been achieved. AgeTech is still in its early stages, and we are not moving fast enough. We need to be honest about where we are falling short. This field demands a level of accountability and critical reflection that we too often avoid.

My challenge to those working in AgeTech is this: stop settling for incremental changes and start pursuing the ideas that feel uncomfortable, risky, or even impossible. Look outside the aging sector for inspiration. The future of AgeTech depends on our ability to think beyond the immediate and envision long-term, systemic change.

If there is one takeaway from this collection, it is that the future of aging is still being written. We have the tools and the knowledge to shape it. The question is whether we're willing to take the necessary risks to make that happen.

Thank you for reading. I hope it was fun. For more fun, check out the additional readings for each essay that are included as the last section of this book.

Recommended Readings for Each Essay

Essay 1: Why Did I Write These Essays? More Importantly, Why Should You Read Them?

1. Topol, E. (2019). Deep Medicine: How Artificial Intelligence Can Make Healthcare Human Again. Basic Books.
2. Achenbaum, W. A. (1995). Crossing Frontiers: Gerontology Emerges as a Science. Cambridge University Press.
3. Gawande, A. (2014). Being Mortal: Medicine and What Matters in the End. Metropolitan Books.
4. McKee, M., & Hunter, D. J. (2020). Public Health in the Age of Populism, Pandemics, and Climate Change. Journal of Public Health, 42(3), 429–431.

Essay 2: If You're Not Aging, You're Dead

1. Lopez-Otin, C., et al. (2013). The Hallmarks of Aging. Cell, 153(6), 1194–1217.
2. United Nations. (2023). World Population Ageing 2023 Highlights.
3. Erikson, E. H. (1998). The Life Cycle Completed. W. W. Norton & Company.
4. Carstensen, L. L. (1992). Social and Emotional Patterns in Adulthood: Support for Socioemotional Selectivity Theory.
5. Kinsella, K., & He, W. (2009). An Aging World: 2008. U.S. Census Bureau.

Essay 3: Addressing the Imbalance: The Urgent Need for Increased Funding in Aging Research

1. Zhavoronkov, A. (2013). The Ageless Generation. Palgrave Macmillan.
2. Fang, E. F., et al. (2020). A research agenda for ageing in China. Ageing Research Reviews, 64, 101174.
3. Olshansky, S. J., et al. (2007). Pursuing the longevity dividend. The Scientist, 21(3), 28–36.
4. Christensen, K., et al. (2009). Ageing populations: the challenges ahead. The Lancet.
5. World Health Organization (2015). World Report on Ageing and Health.

A. Mihailidis, *The Age of AgeTech*, Synthesis Lectures on Technology and Health,
https://doi.org/10.1007/978-3-032-04886-8

Essay 4: Investment in AgeTech

1. Borup, M., et al. (2006). The sociology of expectations in science and technology.
2. van Lente, H., & Rip, A. (1998). The Rise of Membrane Technology.
3. Molotch, H. (2003). Where Stuff Comes From. Routledge.
4. Joyce, K., & Loe, M. (2010). A Socio-Technological Imaginary of Aging in Place.
5. Stirling, A. (2008). Opening Up and Closing Down. Science, Technology, & Human Values.

Essay 5: Public and Private Funding in Aging Research and Services: A Collaborative Imperative

1. Sixsmith, A., & Gutman, G. (2013). Technologies for Active Aging. Springer.
2. Peine, A., & Neven, L. (2021). The Co-Constitution of Aging and Technology. The Gerontologist.
3. Charness, N., & Boot, W. R. (2009). Aging and Information Technology Use.
4. Czaja, S. J., & Lee, C. C. (2007). The impact of aging on access to technology.
5. Neves, B. B., & Amaro, F. (2012). Too old for technology? Journal of Community Informatics.

Essay 6: Sustainable Funding for Aging Services: Securing the Future of Elder Care

1. World Health Organization (2015). World Report on Ageing and Health.
2. National Academies of Sciences, Engineering, and Medicine (2016). Families Caring for an Aging America.
3. AARP & Burning Glass Technologies (2020). The Longevity Economy Outlook.
4. Golant, S. M. (2017). A theoretical model to explain smart technology adoption. Journal of Aging Studies.
5. Mitzner, T. L., et al. (2010). Older Adults Talk Technology. Computers in Human Behavior.

Essay 7: Why AgeTech Matters: Transforming the Future of Aging

1. Coughlin, J. F. (2017). *The Longevity Economy.* PublicAffairs.Selwyn, N. (2004). The information aged. Journal of Aging Studies.
2. Ballantyne, A. (2018). Virtues of Ageing and Care. Bioethics.
3. Vines, J., et al. (2015). Ageing in place: Technology design for rural elderly.
4. Schulz, R., & Czaja, S. J. (2018). The Aging Workforce. American Psychologist

Essay 8: The Rise of AgeTech: Key Innovations, Assistive Devices, and Smart Home Solutions

1. Mitzner, T. L., et al. (2019). Technology Adoption by Older Adults: Findings from the PRISM Trial. The Gerontologist, 59(1), 34–44.
2. Fisk, A. D., et al. (2009). Designing for Older Adults: Principles and Creative Human Factors Approaches. CRC Press.
3. Coughlin, J. F. (2017). The Longevity Economy. PublicAffairs.
4. Neven, L. (2010). 'But obviously not for me': Robots, laboratories and the defiant identity of elder test users. Sociology of Health & Illness, 32(2), 335–347.

Essay 9: Digital Inclusion for Seniors: Bridging the Digital Divide and Designing User-Friendly Technologies

1. Norman, D. A. (2013). The Design of Everyday Things. MIT Press.
2. Pullin, G. (2009). Design Meets Disability. MIT Press.
3. Fisk, A. D., & Rogers, W. A. (2020). Human Factors Interventions for the Health Care of Older Adults. CRC Press.
4. LeRouge, C., & Wickramasinghe, N. (2013). Consumer health informatics: Design, implementation, and adoption. Springer.

Essay 10: AI and Robotics in Elder Care: Companions and Caregivers

1. Tenner, E. (1997). Why Things Bite Back: Technology and the Revenge of Unintended Consequences. Knopf.
2. Greenhalgh, T., et al. (2017). Beyond adoption: A new framework for theorizing and evaluating nonadoption, abandonment, and challenges to scale-up of health and care technologies. Journal of Medical Internet Research, 19(11).

Essay 11: Ethical Considerations in AgeTech: Privacy, Security, and Balancing Automation with Human Touch

1. Loe, M. (2010). Doing it my way: Old women, technology and well-being. Sociology of Health & Illness, 32(2), 319–334.
2. Satchell, C., & Dourish, P. (2009). Beyond the user: Use and non-use in HCI. Proceedings of the 21st Annual Conference of the Australian Computer-Human Interaction Special Interest Group.
3. Joyce, K., & Mamo, L. (2006). Graying the Cyborg: New Directions in Feminist Analyses of Aging, Science and Technology. Studies in Gender and Sexuality, 7(3), 299–324.

4. Vines, J., et al. (2015). Involving older people in design: The role of critical engagement. ACM Transactions on Computer-Human Interaction.

Essay 12: Is Scientific Peer Review Killing Innovation in AgeTech Solutions for Older Adults?

1. Porter, M. E., & Teisberg, E. O. (2006). Redefining Health Care: Creating Value-Based Competition on Results. Harvard Business Press.
2. Taplin, D. H., et al. (2013). Theory of Change Technical Papers: A Series of Papers to Support Development of Theories of Change Based on Practice in the Field.
3. Kamei, T. (2019). Business Models in Aging Societies. In Ageing in the Asia-Pacific Region. Springer.
4. Christensen, C. M., et al. (2009). The Innovator's Prescription: A Disruptive Solution for Health Care. McGraw Hill.

Essay 13: Conducting Research in AgeTech: Paradigms, Approaches, and Challenges

1. Lupton, D. (2016). The Quantified Self: A Sociology of Self-Tracking. Polity.
2. Leslie, D., et al. (2021). Understanding bias in facial recognition technologies. The Alan Turing Institute.
3. Oh, H., et al. (2021). Big Data and Older Adults: Implications for Aging and Technology. Innovation in Aging, 5(2).
4. Hassani, H., et al. (2019). Big data and ageing: A methodological perspective. Journal of Population Ageing, 12(4), 421–439.

Essay 14: Why I Got Into the AgeTech Field: A Personal Journey

1. Winner, L. (1980). Do Artifacts Have Politics? Daedalus, 109(1), 121–136.
2. Eubanks, V. (2018). Automating Inequality: How High-Tech Tools Profile, Police, and Punish the Poor. St. Martin's Press.
3. Joyce, K. (2008). Magnetic Appeal: MRI and the Myth of Transparency. Cornell University Press.
4. Neven, L., & Peine, A. (2017). From intervention to co-constitution: New directions in theorizing about aging and technology. The Gerontologist, 57(2), 247–255.

Essay 15: Should I Go Into a Career in AgeTech?

1. Sherrington, C., et al. (2020). Evidence on fall prevention in older people to inform policy and practice. Public Health Research & Practice, 30(4).
2. World Health Organization. (2017). Global Strategy and Action Plan on Ageing and Health.

3. OECD (2015). Addressing Dementia: The OECD Response.
4. Thomas, S., et al. (2019). Aging Policy and Politics in the US and Abroad. Health Affairs, 38(6), 993–1001.

Essay 16: The Future of AgeTech: Bridging Science Fiction and Reality

1. Christensen, C. M. (1997). The Innovator's Dilemma. Harvard Business School Press.
2. Hart, S. L., & Christensen, C. M. (2002). The Great Leap: Driving Innovation From the Base of the Pyramid. MIT Sloan Management Review.
3. Zuboff, S. (2019). The Age of Surveillance Capitalism. PublicAffairs.
4. West, D. M. (2012). Digital Government: Technology and Public Sector Performance. Princeton University Press.

Essay 17: Interdisciplinary Collaboration: Bridging Gaps to Advance AgeTech Innovation

1. Sharkey, A., & Sharkey, N. (2012). Granny and the robots: Ethical issues in robot care for the elderly. Ethics and Information Technology, 14(1), 27–40.
2. Coeckelbergh, M. (2010). Health care, capabilities, and AI assistive technologies. Ethics and Information Technology, 12(3), 219–231.
3. Borenstein, J., & Pearson, Y. (2010). Robot caregivers: harbingers of expanded freedom for all? Ethics and Information Technology, 12(3), 277–288.
4. van Wynsberghe, A. (2013). Designing robots for care: Care centered value-sensitive design. Science and Engineering Ethics, 19(2), 407–433.

Essay 18: Global Perspectives: Cultural Considerations in AgeTech

1. National Academy of Medicine (2022). Emerging Technologies to Advance Aging Science.
2. Topol, E. (2019). Deep Medicine. Basic Books.
3. European Commission (2020). The Green Paper on Ageing: Fostering Solidarity and Responsibility.
4. Longo, F., et al. (2021). Technological trajectories for aging populations. Futures, 127, 102703.

Essay 19: Education and Training for the Next Generation of AgeTech Professionals

1. Achenbaum, W. A. (2005). Older Americans, Vital Communities. Johns Hopkins University Press.
2. Birren, J. E., & Schroots, J. J. (2001). A History of Geropsychology. Handbook of the Psychology of Aging.

3. Eliopoulos, C. (2017). Gerontological Nursing. Wolters Kluwer Health.

Essay 20: Will We Ever See Anything Novel Again in AgeTech?

1. Barrett, M., et al. (2019). Investing in AgeTech: Opportunities and Risks. Aging Analytics Agency.
2. KPMG (2020). The Silver Economy: Investment and Innovation Opportunities.
3. OECD (2021). Enhancing Access to Quality Long-Term Care Services.

MIX
Papier aus verantwortungsvollen Quellen
Paper from responsible sources
FSC® C105338

If you have any concerns about our products,
you can contact us on
ProductSafety@springernature.com

In case Publisher is established outside the EU,
the EU authorized representative is:
Springer Nature Customer Service Center GmbH
Europaplatz 3, 69115 Heidelberg, Germany

Printed by Libri Plureos GmbH
in Hamburg, Germany